Palgrave Studies in Democracy, Innovation and Entrepreneurship for Growth

Series Editor: Elias G. Carayannis

The central theme of this series is to explore why some areas grow and others stagnate, and to measure the effects and implications in a trans-disciplinary context that takes both historical evolution and geographical location into account. In other words, when, how, and why does the nature and dynamics of a political regime inform and shape the drivers of growth and especially innovation and entrepreneurship? In this socioeconomic and sociotechnical context, how could we best achieve growth, financially and environmentally?

This series aims to address such issues as

- How does technological advance occur, and what are the strategic processes and institutions involved?
- How are new businesses created? To what extent is intellectual property protected?
- Which cultural characteristics serve to promote or impede innovation? In what ways is wealth distributed or concentrated?

These are among the key questions framing policy and strategic decision-making at firm, industry, national, and regional levels.

A primary feature of the series is to consider the dynamics of innovation and entrepreneurship in the context of globalization, with particular respect to emerging markets, such as China, India, Russia, and Latin America. (For example, what are the implications of China's rapid transition from providing low-cost manufacturing and services to becoming an innovation powerhouse? How do the perspectives of history and geography explain this phenomenon?)

Contributions from researchers in a wide variety of fields will connect and relate the relationships and interdependencies among (1) innovation, (2) political regime, and (3) economic and social development. We will consider whether innovation is demonstrated differently across sectors (e.g., health, education, technology) and disciplines (e.g., social sciences, physical sciences), with an emphasis on discovering emerging patterns, factors, triggers, catalysts, and accelerators to innovation, and their impact on future research, practice, and policy.

This series will delve into what are the sustainable and sufficient growth mechanisms for the foreseeable future for developed, knowledge-based economies and societies (such as the EU and the United States) in the context of multiple, concurrent, and interconnected "tipping-point" effects with short-term (MENA) as well as long-term (China, India) effects from a geo-strategic, geo-economic, geo-political, and geo-technological set of perspectives.

This conceptualization lies at the heart of the series, and offers to explore the correlation between democracy, innovation, and growth.

Books Appearing in this Series:

Unpacking Open Innovation: Highlights from a Co-Evolutionary Inquiry, Manlio Del Giudice, Elias G. Carayannis, and Maria Rosaria Della Peruta

Unpacking Open Innovation

Highlights From a Co-Evolutionary Inquiry

Elias G. Carayannis, Manlio Del Giudice, and Maria Rosaria Della Peruta

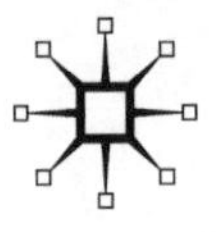

UNPACKING OPEN INNOVATION

First published in 2013 by
PALGRAVE MACMILLAN®
in the United States—a division of St. Martin's Press LLC,
175 Fifth Avenue, New York, NY 10010.

Where this book is distributed in the UK, Europe and the rest of the World, this is by Palgrave Macmillan, a division of Macmillan Publishers Limited, registered in England, company number 785998, of Houndmills, Basingstoke, Hampshire RG21 6XS.

Palgrave Macmillan is the global academic imprint of the above companies and has companies and representatives throughout the world.

ISBN: 978–1–137–35932–2

Library of Congress Cataloging-in-Publication Data

Carayannis, Elias G.
Unpacking open innovation : highlights from a co-evolutionary inquiry / Elias G. Carayannis, Manlio Del Giudice and Maria Rosaria Della Peruta.
pages cm
Includes bibliographical references and index.
ISBN 978–1–137–35932–2 (hardback)
1. Technological innovations. 2. Diffusion of innovations.
3. Research, Industrial. I. Del Giudice, Manlio.
II. Della Peruta, Maria Rosaria. III. Title.
HC79.T4C3684 2014
338′.064—dc23 2013026831

A catalogue record of the book is available from the British Library.

Design by Integra Software Services

First edition: December 2013

10 9 8 7 6 5 4 3 2 1

Contents

List of Figures vii

Introduction ix

1 Open Innovation in Management Science 1
Elias G. Carayannis

2 Open versus Closed Innovation: Speculating about the Future of Technology Management 27
Elias G. Carayannis

3 Invention, Inventiveness, and Open Innovation 51
Manlio Del Giudice

4 Open Innovation or Collective Invention? Conceptualizing the Debate 69
Manlio Del Giudice

5 Openness That Matters: Net Generation, Higher Education, and Student Entrepreneurship 91
Manlio Del Giudice

6 On the External Dimension of Business Knowledge Flows: "Markets for Knowledge Resources" 119
Maria Rosaria Della Peruta

7 What Open Innovation Is: Local Search, Technological Boundaries and Sustainable Performance in Biopharmaceutical Experimentation 155
Maria Rosaria Della Peruta

Index 185

List of Figures

1.1	Going beyond to the innovation	2
1.2	On processes of innovation	4
1.3	The emergence of OSS communities	5
1.4	Knowledge sourcing schemes and innovation	6
1.5	Technological and user knowledge	17
2.1	Innovation inside and outside the company	28
2.2	The Chinese innovation system	30
2.3	Open source initiatives	32
2.4	Peer production and collective creativity	35
2.5	Innovation and communities	38
5.1	Local versus global informational advantages	96
6.1	Markets of knowledge resources	120
7.1	European Lead Factory: An Open Innovation Experiment in Drug Discovery	166

Introduction

There is quite a confusion with regard to the concept of open innovation, so it is necessary to thoroughly analyze all the different approaches to the field. The relevant literature has utilized a wide range of terms, such as distributed innovation, free innovation, collegial innovation, collaborative innovation, free knowledge disclosure, and open knowledge disclosure, just to mention the most common. At this point, the relevant question is: do all these terms have the same meaning? For a more complete understanding, it is therefore essential to shed light on all the possible implications of the concept of open innovation. However, first of all, it is important to explain the popularity of open innovation in the recent decades, and to do so it is necessary to analyze the typical configuration of the R&D and commercialization process in most companies. The conventional idea is of a laboratory in which scientists do their research, while managers are involved in development and commercialization. The routine R&D and commercialization process discloses some relevant issues. It is quite improbable that all the best ideas are born within a single company, and even considering a single idea, it is not conceivable that all the technical and market risks related to the commercialization process should be handled by one organization. This is when open innovation applies. In Chesbrough's own words, open innovation can be described as "a paradigm that assumes that firms can and should use external ideas as well as internal ideas, and internal and external paths to market as the firms look to advance their technology." As most scholars believe, through open innovation companies are enabled to extend their range of ideas and opportunities. At the same time, the technical and market risks related to innovation are reduced. Moreover, it seems that there are almost no drawbacks ensuing from open innovation. This book aims at investigating the research field of open innovation at a stage that is constantly evolving. The most significant topics in literature are identified in order to highlight the most relevant trends of publications. Another purpose is to analyze in depth some fields of inquiry that could prove to be significant but have not received due attention so far.

If the concept of open innovation is enlarged just a little to incorporate other innovation processes that are carried out by companies or individuals beyond the normal boundaries of firms engaged in research and development, it is possible to ascertain that open innovation has often been accomplished in one way or another by most organizations. A different set of nonconventional innovation processes is usually implemented, which could quite closely resemble open innovation. Although many of these processes cannot be considered routine R&D processes, they are commonly utilized by companies to carry out innovation initiatives. This means that the open innovation process is much wider than suggested by a few relevant indicators. It is therefore fundamental to understand what occurs in such a process, and examine in depth the business models and mechanisms that are often related to open innovation as Chesbrough defines it. To achieve this goal it is necessary to investigate the different types and purposes of open innovation, and develop at the same time conceptual progress and empirical knowledge creation. The field is dynamic and constantly evolving, as proven by the fact that a great number of companies are involved in projects that are open, although not in line with the standard definition of open innovation, and embrace work on knowledge that tends continuously to extend boundaries. For an empirical specification of the evolving characteristics of open innovation processes, a higher level of intellectual activity would have been required. Nevertheless, it has been sufficient to elaborate a detailed analytical framework to guide future researchers.

THEORETICAL ADVANCES IN OPEN INNOVATION

The term *open innovation* is still the subject of much debate, but most scholars believe that the concept is liable to a wider field of application compared to Chesbrough's vision. Today, open innovation can be considered a paradigm that links research from different areas of management sciences. The common goal is understanding how a company can be more innovative, and to reach this aim researchers are expanding in multiple directions and debating on a great variety of fields.

OPEN VERSUS CLOSED KNOWLEDGE DISCLOSURE

The model of *closed innovation* was based on self-reliance, according to which innovation processes should be always controlled by the firm. Over time, the boundaries to the innovation processes began to extend, due to significant changes in industry and society. New financial structures, such as venture capital, have appeared, and knowledge workers have become more mobile. This has led to an *open innovation* model according to which both external and internal resources are utilized by companies, and both external and internal technologies are commercialized. In an open innovation process, a project can ensue from an internal or an external source, and new technology can apply at any stage of the process. Also the ways the new ideas can reach the market may vary; besides the classical sales channels, spin-off ventures and out-licensing are an option.

OPEN INNOVATION OR COLLECTIVE INVENTION?

Recently, there has been an intense debate on the significance of knowledge and knowledge management for firms. From a historical standpoint, most organizational knowledge was created inside companies, for example, by a firm's research and development department. Nowadays, only a small number of organizations can base themselves on internal knowledge sources to continue being innovative and competitive. If they are willing to seize the most recent tendencies, they have to constantly gain knowledge from outside. The key to innovation is an effective management of highly specific knowledge from customers, markets, and other sources. Its significance is shown by high interest in researching on topics such as collective invention or user innovation, just to mention a couple of examples. Nevertheless, numerous difficulties arise when organizations integrate external knowledge in order to encourage innovation. A stimulating approach in current research is considered the analysis of open innovation as a paradigm shift in the strategic approach incorporating the external world into internal innovation processes.

OPEN SCIENCE VERSUS SCIENCE OPEN: THE EMERGING ISSUE

Today the word "open" has to be placed before the word "science" in order to distinguish open science from all the other types covered by intellectual property (IP) restrictions so frequently carried out by academic institutions and private research organizations. It is a shame that this has been the outcome, since science requires to be open and transparent, as it regards reproducible results under controlled conditions. Luckily, there is no more alarm and many researchers are now handling the matter. There are three most common ways in which

IP barriers in science are established. First, the traditional publishing process is less able to sustain reproducibility. When something is published without detailing relevant data and methods, and without a valid documentation supporting it, reproducibility is almost impossible and many could even question what is being shared. Some papers may be so complex that a long time is required to reproduce the science behind them, and many researchers may even decide not to take that path because of the pressure to generate new ideas in a short time. In fact, it is confirmed by evidence that those who create something new are more well recognized than those who merely confirm a result. Second, at times the open research process faces fierce opposition by some researchers who do not desire that others see the innovation process exposed. Nowadays, science is a real competitive setting, and many researchers prefer that their ideas are protected, also considering the issues related to obtaining research funding. Third, at universities and other research institutions, formal IP organizations have been created in order to control and license IP. These barriers are being addressed in different ways thanks to the increased importance of networked science and computing. These new methods support open science (community) by providing open access (publication), open source (software), and open data. *PLoS* and *BioMed Central* are examples of innovative open access journals. Even more amazing are those journals that receive software and submit it to human reviewers only after testing and scoring it automatically. Following the increasing success of computing, it is fundamental that the path of innovation is maintained clear in order to open up the way to an enlightened future. Open science plays a key role in this, and although many fear the opposite, commercialization is not prevented by such a practice. In the years to come, impact factors should not be the sole measures of success; also the ways sharing and re-usage of research outcomes occur should be considered, since the main purpose should be to reach results upon which others can build.

Co-evolution of Open Innovation and Corporate Innovation

Open innovation can be defined as the formal discipline and practice of exploiting something new and not obvious discovered by others and using it in the innovation process through the establishment of formal and informal relationships. Some of the major topics and issues to be dealt with are as follows:

- Why open innovation capacities and tools are deployed by companies?
- What important considerations can be drawn from the cases of General Motors, Philips, and B/S/H?
- What is the role of open innovation in the current economic context?
- How can open innovation be measured?
- What is the role of open innovation among the innovation tools companies have at their disposal?

Here, experiences regarding implementation and execution of open innovation are examined. It is shown how to turn open innovation into reality, by applying methods and practices successfully. Also the role played by open innovation within the different tools used by firms is analyzed, together with the synergies implied, and the way a global innovation strategy is driven in such a context.

KNOWLEDGE, INNOVATION, AND ENTREPRENEURSHIP: HOW THE RELATIONSHIP CAN BE RECONSIDERED IN TERMS OF OPENNESS

The knowledge-driven economy affects the innovation process and the approach to innovation. The traditional idea that innovation is based upon research (technology-push theory) and interaction between firms and other actors is replaced by the current social network theory of innovation, where knowledge plays a crucial role in fostering innovation. The most interesting outcome is that both network structure and access to heterogeneous knowledge are equally important for general organizational performance, and even more significant for innovation performance. What assumptions about the social and the technical are embedded in the existing strands of literature? What are the strengths and weaknesses of the different theoretical perspectives and methodological approaches in terms of conducting research and understanding social media? How can the theories be applied to and/or generated from empirical studies of real-world cases of social media engagement in organizational settings? How can methods and tools for social media engagement, analytics, and management be designed, developed, and evaluated and what are the managerial implications and societal consequences? Based on a growing empirical literature, we can begin to find patterns consistent with fundamental theoretical formulations.

While open innovation has reached high popularity as a tool to increase innovation in companies, the concept has not been as well understood with regard to entrepreneurship. Here the role of entrepreneurship in the open innovation process is thoroughly investigated. In innovation-related research, problems associated with open innovation and entrepreneurship have been a matter of debate. However, the relationship between the two has not been properly analyzed and is far from being completely understood. It is believed that different open innovation channels are utilized by the different entrepreneurship practices; as a result, the impact on open innovation varies. Hereafter, the relationship between open innovation and entrepreneurship will be investigated with regard to the IT convergence industry sector in emerging and developed economies.

Chapter 1

Open Innovation in Management Science

Elias G. Carayannis

> *There are different sources of open innovation. A classical one is knowledge spillovers, which arise when firms can capture knowledge or information "in the air," as Marshall put it. Recently, there has been an upsurge in the so-called "open source" phenomenon whereby knowledge and information are distributed openly by their producers, in a context where the production and distribution of knowledge are governed by well-defined norms (e.g., Lerner and Tirole, 2002). An "old" form of open source is open science, which is again based on clear norms of production and diffusion of knowledge (Dasgupta and David, 1994). Open science, and particularly the proximity of firms to universities or other scientific institutions, have themselves been considered sources of knowledge spillovers (e.g., Alcacer and Chung, 2007).*
>
> (Gambardella, 2010, p. 85)

The concept of open innovation was introduced by UC Berkeley professor Henry Chesbrough, who has become famous worldwide due to his work *Open Innovation—The New Imperative for Creating and Profiting from Technology*, which was published in 2003. The scholar showed how, in the past century, highly innovative ideas were generated by firms that strongly invested in internal research and development and hired top professionals. These ideas were protected by an effective intellectual property (IP) strategy. Generally, a virtuous

In ***the old model of closed innovation***, enterprises adhered to the following philosophy: Successful innovation requires control. In other words, companies must generate their own ideas, then develop, manufacture, market, distribute and service those ideas themselves. For most of the 20th century, that model worked well, as evidenced by the spectacular successes of central R&D organizations such as AT&T's Bell Labs.

Today, though, ***the internally oriented, centralized approach to R&D is becoming obsolete in many industries***. Useful knowledge is widely disseminated, and ideas must be used with alacrity. If not, they will be lost. Such factors create ***a new logic of open innovation***, in which the role of R&D extends far beyond the boundaries of the enterprise. Specifically, companies must now harness outside ideas to advance their own businesses while leveraging their internal ideas outside their current operations.

H. W. Chesbrough (2003b) "Open Innovation: The New Imperative for Creating and Profiting from Technology"

Figure 1.1 Going beyond to the innovation

circle of innovation was activated, since profit was reinvested in R&D (Chesbrough, 2003a) (figure 1.1).

Nevertheless, in the last years of the twentieth century, innovation management changed due to several reasons, in particular, (1) the number of knowledge workers increased together with their mobility and (2) venture capital became increasingly available. As a result, the closed innovation process in companies started to fall apart (Chesbrough, 2003a).

Some other reasons pointed out by Chesbrough (2003b) are as follows:

- Widespread circulation of useful knowledge
- Firms' inadequate exploitation of available information
- Loss of ideas that are not immediately used
- An unsupportive business model, on which the importance of an idea or a technology depends
- Alteration of the innovation process by the presence of venture capital
- The need for firms to be active sellers and buyers of IP

Hence, an *open innovation* model was developed, according to which companies commercialize both external and internal ideas by implementing specific routes that lead to and from the market. In Chesbrough's words (2003a, p. 37), "the boundary between a firm and its surrounding environment is more porous, enabling innovation to move easily between the two." In a similar process, projects can ensue from internal or external sources and new technology may be deployed in different phases. Moreover, the ways projects reach the market can vary; besides the classical sales channels, it is possible to start up a spinoff venture or resort to out-licensing (Chesbrough, 2003b).

Principles of Open Innovation

According to Chesbrough (2003a), open innovation is founded on the following principles:

- Intelligent people do not all work in-house and therefore external knowledge is required.
- Significant value may be created by external R&D.
- Research profitability is not exclusively associated with internal work.
- "First to market" is less important than a strong business model.
- Both external and internal ideas are fundamental to win.
- For a company it is important to capitalize on its own IP and purchase external IP when necessary (figure 1.2).

West, Vanhaverbeke, and Chesbrough (2006, p. 286) defined open innovation as "both a set of practices for profiting from innovation and also a cognitive model for creating, interpreting and researching those practices." Chesbrough (2006a) has argued that two anomalies in previous research on innovation have been overcome by open innovation. First, the spillovers are no longer considered as something to avoid but as a direct result of the business model; second, the IP rights are no longer deemed as a tool for protection but as a new class of assets. The current business model may benefit from both.

Moreover, five key themes in research were identified (Chesbrough, 2006b):

- The business model—value is generated within the value chain and captured in part by focal firms due to its two important functions, discussed in the next section.
- External technologies—they fill the gaps and enable the manufacture of complementary products that allow technology to be more rapidly accepted, so the company's business model is leveraged.

- Complexity of knowledge identification, evaluation, and incorporation—knowledge management and connections appear increasingly relevant.
- Start-ups—they represent experiments with business models, as they carry new technologies and allow exploration of new markets.
- IP rights—by this means ideas and technologies can be easily transferred.

In the industry of open source software development (OSS), some open innovation models were initially developed and later transferred to more general open innovation practices. An OSS project involves a decentralized community of volunteer developers who collaborate

"Open Innovation follows a long tradition of studying the processes of innovation, and stands on the shoulders of many previous scholars" (Chesbrough 2006a, p. 5).

As Pavitt (2002, pp. 119–120) emphasized:

So far, this writer has been unable to find a simple or elegant theoretical framework to encompass the richness of the empirical material on corporate innovative activities. However, in organising this material, it has proved useful to divide the processes of innovation into three, partially overlapping, processes each of which is more closely associated with contributions from particular academic disciplines.

- Producing scientific and technological knowledge: since the industrial revolution, the production of scientific and technological knowledge has been increasingly specialised, by discipline, by function and by institution. Here, history and social studies of science and technology have been the major academic fields contributing to our understanding
- Transforming knowledge into working artefacts: in spite of the explosive growth in scientific knowledge in this period, theory remains an insufficient guide to technological practice, given the growing complexity of technological artefacts, and of their links to various fields of knowledge. Technological and business history has made major contributions here and, more recently, so have the cognitive sciences.
- Matching working artefacts with users' requirements: the nature and extent of the opportunities to transform technological knowledge into useful artefacts vary amongst fields and over time, and determine in part the nature of products, users and methods of production. In the competitive capitalist system, corporate technological and organizational practices therefore co-evolve. These processes are central concerns of scholars in management and economics.

Figure 1.2 On processes of innovation

OSS communities are open in the sense that their outputs can be used by anyone (within the limits of the license) and anyone can join merely by subscribing to an e-mail list. Openness in terms of membership leads, in turn, to transparency in the development process, since communication about projects and their direction largely occurs in public. Thus, project leadership is accountable to the wider community for its growth and future direction, and everyone is aware of shortfalls and issues. Transparency also affords individuals self-determination with respect to the level of effort they choose to expend and awareness of others' efforts that they might be able to fold into their own (Gulati, Puranam and Tushman, 2012, p. 579).

Figure 1.3 The emergence of OSS communities

to produce a software product using Internet-based tools such as e-mail, mailing lists, Web-based concurrent versioning systems (CVS), and bug-reporting software. To date, OSS researchers and practitioners have been primarily interested in three sub-areas of research: (1) individual developer participation; (2) competitive dynamics; and (3) innovation processes, governance, and organization [see von Krogh and von Hippel (2006) for a summary of these areas] (figure 1.3).

According to West and Gallagher (2006), three fundamental challenges of open innovation may be identified: motivation, integration, and exploitation of innovation. These were examined by the two scholars through a qualitative and quantitative analysis of OSS development.

Four generic open innovation strategies were detected:

- Pooled R&D and shared R&D, for which a cultural change is required
- Spinouts, which are a means of bypassing large companies' bureaucracies
- Selling complements, which means agreeing to commoditization or developing different types of products according to different commodities
- Using donated complements, which means that differentiated products can be developed by users thanks to the availability of general-purpose technologies

The open innovation model is becoming widely adopted in many industries nowadays. For example, in the pharmaceutical industry, open innovation has become one of the most significant trends. In the area of pharmaceutical research, Henderson and Cockburn (1994,

p. 67) have shown that the ability to "encourage and maintain an extensive flow of information across the boundaries of the firm" is important to the productivity of the process of drug discovery. In fact, pharmaceutical firms have realized that it is too expensive to have all competences in-house, so they have begun to focus on the most important, including technology platforms and therapy areas, and collaborating at the same time with the right partners.

In their study on pharmaceutical firms, Bierly and Chakrabarti (1996) uncovered four generic knowledge strategy groups—"explorers," "exploiters," "loners," and "innovators"—and found that firms with a good balance of both internal and external learning with a tendency toward more radical learning (i.e., "innovators" and "explorers") exhibited consistently higher levels of profitability (figure 1.4).

Different interesting implications result from the issue of whether a company gains its knowledge from sources inside or outside the organization, and how the different acquisition behaviours affect the propensity to innovate. A number of researchers have analyzed the knowledge sourcing schemes implemented by companies, reaching the conclusion that both internal and external sources of knowledge are on the same level of importance. Iansiti and Clark (1994) studied the automobile and mainframe computer industries, finding that considerably high performing companies were actively involved in internal and external integration. In their research on the lithotripsy industry, Nagarajan and Mitchell (1998) revealed that the methods of knowledge acquisition were responsible for different types of technological shifts. This means that for encompassing and complementary changes the companies relied on inter-organizational relationships, while for incremental shifts they trusted in internal Research and Development. Nobel and Birkinshaw (1998) analyzed communication and control in international R&D operations, revealing that international creators, who were more responsible for innovation than improvement and adjustment, preserved effective internal and external networks of relationships. This means that they were able to acquire knowledge from external sources, and implement coordination and communication across organizational sub-units. In their study on the retail food industry, Rulke et al. (2000) ascertained that store managers based their organizational self-knowledge on both internal and external sources of information. In their research on the optical disk industry, Rosenkopf and Nerkar (2001) discovered that technological developments were highly influenced by explorations extended along both organizational and technological boundaries.

Figure 1.4 Knowledge sourcing schemes and innovation

Source: Adapted from Pedersen, T., Soo, C. & Devinney, T. M. 2011, 'The Importance of Internal and External Knowledge Sourcing and Firm Performance: A Latent Class Estimation'. in C. Geisler Asmussen, T. Pedersen, T. M. Devinney & L. Tihanyi (eds), *Dynamics of Globalization: Location-Specific Advantages or Liabilities of Foreignness?*. Emerald Group Publishing Limited, Bingley, pp. 389–423. *Advances in International Management*, vol. 24.

As pointed out by Gassmann and Reepmeyer (2005), one of the most important goals in R&D management today is balancing the right size and structure. In the biotechnology industry, "research breakthroughs demand a range of intellectual and scientific skills that far exceed the capabilities of any single organization" (Powell et al., 1996, p. 118). Various studies on R&D management (Tushman and Katz, 1980; Ebadi and Utterback, 1984) have indicated that in a dynamic technology-intensive research environment, the ability to span organizational boundaries is extremely important.

Moreover, Gaule (2006) widely based his analysis on Chesbrough (2003b) and his own consulting model for open innovation in order to show how many parts of an organization are influenced by the latter. He also examined some short case studies, such as the case of Procter & Gamble. Motzek (2007) highlighted the motivation factors that lead firms to adopt an open innovation model. In particular, he takes into account two new organizations that embrace open innovation right from the start. Here the motivation factors are quite similar to the motivation factors for entrepreneurs, so there is a difference compared with pre-existent firms that have to change in order to engage in open innovation.

A study by Gruber and Henkel (2006) revealed that in the case of open source software, the fundamental challenges for new companies discussed in the entrepreneurship literature are less important. These factors can be summarized as follows:

- Newness—when companies participate actively, they become well known and it is possible for them to quickly create a public track record.
- Smallness—market offering based on code that is freely accessible and informal collaborations are advantages.
- Market entry barriers—the company benefits from earlier development endeavors, rapid focus on differentiations, and low switching costs for users.

Gruber and Henkel (2006) also stated that there is the possibility of a more general application of open source software to any industry where blueprints can be exchanged online and innovation proceeds gradually.

An interesting metaphor was used by Chesbrough (2004) to describe innovation management. The author referred to chess and poker and claimed that when the number of sources in a firm increases, it is extremely important to assess early-stage technologies. When a known market is targeted with a new technology, all information is known; it is like playing chess—everyone knows the pieces and how they can move. Instead, when there is a situation in which both technology and markets are new, the ensuing path is not only unknown but also *unknowable*. The probabilities of measurement errors, such as false positives and false negatives, are high. False positives are ideas or technologies that are deemed highly potential but turn out to be less successful than expected, whereas false negatives are ideas or technologies that are mistakenly underestimated. The management of false negatives can be better understood through the metaphor of playing poker. As stated by Chesbrough (2004), to play poker, firms "need to measure their capital and stage their investments in projects upon the receipt of new information." Poker strategies involve the following:

- Overview of what occurs inside after the decision to no longer resort to financing
- Revelation of the failure outside so new perspectives can be foreseen
- Out-licensing of the rejected project
- Creation of a spinoff venture

In chess the objective to find a point of convergence between the pathway of the future project and the current business model must achieve a net present value greater than 0 and minimize false positives. In poker the aim is to create options for future business models, leverage or increase business, achieve an options value greater than 0, and manage false negatives (Chesbrough, 2004).

Open Business Models and Value Creation

According to Chesbrough (2007), there are two functions of a business model: the creation of value and the capture of part of that value. Open business models enhance the efficiency of firms in such a task. Furthermore, firms must adjust their business models to open innovation, because it is a means of creating value from their own IP (Chesbrough, 2003c). Moreover, firms find it increasingly difficult to justify investments in innovation due to the increasing costs of developing technology and the reduced life cycles of their products. By resorting to an open business model, a firm can reduce costs and save time by leveraging on external R&D resources and, at the same time, licensing out internal technologies in order to increase revenues. Chesbrough (2007) also believes that firms should extend their abilities by experimenting with their business models, for example venturing in alternative brands or spinoffs in order to cut down risks. These are all significant changes for which intense commitment and support are required.

As pointed out by Chesbrough and Schwartz (2007), in open innovation models, co-development partnerships are becoming more and more important. This is consistent with Chiaromonte's view (2006) that the difference between open innovation and classical outsourcing of innovative capabilities is that the external partners are deemed as peers and not as mere suppliers.

Chesbrough and Schwartz (2007) argue that R&D costs can be reduced by business models based on the use of partners, and, at the same time, innovation can be expanded and new markets opened.

In order to achieve this goal, the two scholars stress the need to

- Identify precisely the business goals for partnering
- Distinguish the company's R&D capacities as follows:
  - core (key source of advantage)
  - critical (essential for success but not key)
  - contextual (required to complete the offering, but not a differentiator)
- Align the business models of the two companies

A research by Van der Meer (2007) conducted in the Netherlands revealed that an articulated business model was present only in a limited number of firms, and this did not allow them to be adequately flexible, for example, through successful implementations that went beyond the original field of business.

Organizational Mechanisms and Boundaries of the Firm

Chesbrough (2003b) argued that complete openness is not practised by all firms. In most cases, there is a sort of continuum between different levels of openness. Moreover, in their open innovation models, firms can have a set of different roles (Chesbrough, 2003b, p. 13).

Hence, it is possible to distinguish between different types of organizations, as follows:

Organizations that finance innovation:

- Innovation investors (such as incubators, private equity, venture capital) and innovation benefactors (early financing)

Organizations that create innovation:

- Innovation explorers (that discover new research functions)
- Innovation merchants (that codify and trade IP)
- Innovation architects (that generate value by building a system that enables single pieces to be brought together)
- Innovation missionaries (that have in mind a cause and try to serve it by creating technologies and making them progress)

Organizations that bring innovation to the market:

- Innovation marketers (that successfully market new ideas)
- Innovation one-stop centers (that sell the ideas of other companies)

Furthermore, there are organizations that still try to maintain control over all parts and are described by Chesbrough (2003b) as "fully integrated innovators."

Jacobides and Billinger (2006) argued that the scope of a company and its degree of openness to final and intermediate markets are identifiable through its vertical architecture. Vertical architectures that are open in part to the markets along the value chain are described as permeable. If the level of permeability is high, resources can be utilized more efficiently, market requirements and capacities are identified more easily, and the creation of more open innovation platforms is encouraged. The authors discussed a case study of a fashion company that revealed how increased permeability could induce successful change in the vertical structure, and this could lead to better strategic and productive capacities, enhanced innovation potential, and improvements in the resource allocation processes (Jacobides and Billinger, 2006). Consistent with this view, Tao and Magnotta (2006) described the sourcing process at Air Chemicals, where the establishment of a wider interface toward various sources of knowledge at a global scale was attempted. Moreover, Fetterhoff and Voelkel (2006) highlighted the issues regarding the search process for innovations. When the authors describe innovation, they focus on the merging process between customer demands and technology, since the former needs to be satisfied by the latter. In fact, companies have to deal with a series of problems when managing external innovations, since, in most cases, they are not used to assess them (Fetterhoff and Voelkel, 2006). These issues can be summarized as follows:

- Searching for opportunities
- Assessing the market potential and inventiveness of an opportunity

- Recruiting potential partners trying to convince them with effective arguments
- Seizing value through commercialization
- Expanding innovation provision together with an external partner

With this line of thought, it is inevitable to mention a case study by Dittrich and Duysters (2007) on Nokia's external contacts. Initially, Nokia had developed its products internally, while for its third-generation mobiles, the company decided to collaborate with external firms, thus opening up its business. Although at first Nokia had had long-term partnerships based on the exploitation of innovations, now the firm has started explorative collaboration agreements extended to companies with which its relations were weaker. Earlier "strong tie" exploitation agreements were founded on immovable structures, while the current new model based on "weak ties" meant a much more organic way of working. Also Simard and West (2006) differentiated the kinds of ties that link firms together. It is possible to distinguish *deep ties* that allow a company to capitalize on existing knowledge and resources from *wide ties* that allow a firm to discover new technologies and markets. Both deep and wide ties can be found in open innovation networks, and they can be formal or informal. Moreover, the scholars stated that incremental innovations are the typical outcome of *deep networks* (Simard and West, 2006).

Furthermore, Dahlander and Wallin (2006) attempted to find the solution to the issue regarding the way companies can use communities as complementary assets, although they do not own them or control them hierarchically. Communities protect their own work through practices they have developed, and companies have to assign individuals to work in these communities, if they only want to access them or even influence them. Sponsored individuals tend to interact more with the rest of the community, and especially with those in the leading positions. Brown and Hagel (2006) argued about the new phenomenon of creation nets in which a number of people directed by a network organizer or gatekeeper work together in order to generate new knowledge, learn from one another, and make progress based on everyone's contribution. One of the best examples provided is that of Linux.

The gatekeeper starts up a creation net by choosing the participants and specifying the participation protocols. Usually there is complete freedom within well-identified interfaces, as modular processes to structure the activities are defined. Moreover, action points are specified by creation nets that state the time of delivery. Participants have to be motivated, so a set of long-term incentives has to be offered. New management approaches are also necessary to

- Decide which is the best way to coordinate the network
- Balance local innovation with global integration
- Specify effective action points
- Create useful performance feedback loops

The scholars believe that creation networks are the best solution when demand for goods and services is not certain, when several different specialists is required in order to make innovation possible, and when there are rapid changes in performance requirements.

Lichtenthaler and Ernst (2006) introduced new ideas for organizational attitudes in the management of boundaries. In an open innovation context, knowledge transactions with the external setting have to be organized by managers. When managing knowledge, three fundamental tasks are involved: knowledge acquisition ("make or buy"), knowledge integration ("integrate or relate"), and knowledge exploitation ("keep or sell").

The authors based their argument on the classical "not-invented-here" (NIH) syndrome, adding that with regard to the external organization of knowledge management, an overly positive attitude is possible. Thus, six distinct syndromes can be identified, and they can be overly positive or overly negative to every single process:

- Make or buy: negative, NIH; positive, buy-in (BI)
- Integrate or relate: negative, all-stored-here (ASH); —positive, relate-out (RO)
- Keep or sell: negative, only-use-here (OUH); positive: sell-out (SO)

As more open innovation systems are adopted, knowledge assets will have to be increasingly commercialized externally by firms, if these wish to keep up with competitors. According to Lichtenthaler (2007a), it is possible to identify three important factors that can help companies reach strategic fit in the choice to keep or sell: coordination, centralization, and collaboration. First, companies should deem the exploitation of external knowledge as a core activity. Second, coordination with other strategies is required together with centralization, that is, providing a definite direction. Third, in order to minimize interface issues, cross-functional collaboration is essential.

Technology licensing is a means to exploit knowledge assets. Lichtenthaler (2007b) believes that technology licensing usually ensues from a combination of diverse drivers and motivation factors. The author surveyed a number of medium-sized and large industrial companies identifying two main drivers, the first of which is *ensuring freedom to operate* and the second is *gaining access to another company's technology portfolio.* The monetary dimension did not become as significant as initially assumed.

Entrepreneurial Models and Innovativeness

"The entrepreneur is the leader who leads the firm to new techniques" in Schumpeter's terms, the one who "carries out new combinations." He gives five interpretations, covering (i) new products, (ii) new means of production, (iii) new markets, (iv) new sources of supply, and (v) new organization of industry (e.g., creating a monopoly). The general emphasis in Schumpeter's work is, of course, on the entrepreneurial phenomenon in its most pure and dramatic form, where single individuals provide the leadership needed to bring about drastically new ways of doing things. (Winter, 2006, p. 137)

Most papers draw the conclusion that it is important that people who endeavor to be innovative are supported by leadership (Carayannis et al., 2007). Nevertheless, only a few examine leadership in open innovation. Fleming and Waguespack (2007), in particular, focused their analysis on leadership in open innovation communities. The scholars argued that valid technical contributions must be made by the future open innovation leader from a structural position able to bind the community together. This is in line with the typical norms of an engineering culture and is made possible by two related but different social positions: social brokerage and boundary spanning between technological fields. Physical interaction may allow to bypass an inherent absence of confidence related to brokerage positions. Boundary spanners, instead, have higher probabilities of becoming leaders since they do not undergo this handicap.

Intensive research on careers within the Internet Engineering Task Force community, which is the first open innovation community in the world, in the period 1986–2002, validates such a statement. According to Witzeman et al. (2006), not only the technological systems require to be transformed. Enterprise systems, processes, values, and culture also require more changes as companies increasingly resort to external innovation. The firms included in the sample resisted open innovation. The search for new external technologies was hindered by strong internal forces that preferred relying upon current in-house technology. Witzeman et al. do not think that this is anomalous, because individuals who work for firms are usually trained to think internally, and concepts such as core competences and Six Sigma reinforce this trend. Those who are capable of involving external sources in their procedures become the leaders, and this is truly challenging. In the words of Witzeman et al. (2006),

> Building external thinking into the firm requires change. The firm must review the new product development processes, the supply chain, the strategic planning process, the reward system, the technology roadmap, and many other systems for their ability to incorporate external innovation.... Harnessing external technology for innovation requires a fundamental change in employee thinking. The "Not Invented Here" syndrome is replaced with the "Invented Anywhere" approach.

This is also consistent with arguments by Dodgson, Gann and Salter (2006). who believe that (1) both cultural changes and new skills are needed, (2) current practices are not replaced, and (3) the uncertainty of innovation is not overtaken by technology.

Tools, Technologies, and Processes Enabling Open Innovation

Many studies debate on the technological interface that allows companies to collaborate with a high number of customers. Dodgson, Gann and Salter (2006) claimed that changes in the technological interface require a transformation in the capacity of an organization to evaluate or assimilate external impressions, and this demands a specific preparation. The authors analyzed the case study of Procter & Gamble; the multinational company developed a technique known as "Connect & Develop" (C&D), a means of internal connection aimed at improving the attitude for initiatives originated externally (see also Huston and Sakkab, 2006, 2007).

The technologies, tools, and processes that enable open innovation may be generally distinguished, as discussed in the following subsections.

Coordinating/Aggregating

It is possible here to refer, for instance, to the Procter & Gamble C&D model that is utilized to leverage sources internally and externally in order to exploit the distributed innovative capability (Dodgson, Gann and Salter, 2006; Huston and Sakkab, 2006, 2007). This model avails itself of the wide interface of a multinational company toward external parties at a global scale to unveil ideas for new products, comprehend customer demands, and discover solutions to technical issues. In the same line, Tao and Magnotta (2006) analyzed the process known as "Identify and Accelerate" (I&A), which is utilized to better understand the peculiar requirements of a company and expand its interface toward the market through cooperation with external search providers in order to meet those demands. Standard open-source methods (see Henkel, 2006), and the toolkits utilized for innovation and mass customization (Piller and Walcher, 2006), may also be described as coordinating/aggregating.

Liberating

According to Piller and Walcher (2006), customer knowledge is "sticky" and traditional market research does not easily unveil it. They believe that contests could be a way to encourage customers to be creative in order to unearth hidden knowledge and preferences, and exploit them.

Allowing/Including

As much research suggests, many issues concern the construction of the structures that should sustain the use of open innovation. One of these issues is that current models mainly focus on internal sources of ideas and competence, rather than external ones. A shift in culture and behavior is therefore required, and in order to achieve this goal the formal models that direct the work process should be the starting point of such a transformation. According to Gassmann, Sandmeier and Wecht (2006), the innovation process can be opened up by utilizing the extreme programming (XP) software development model. Customer interaction is enhanced by the iterative features of the process, and XP may be applied to the development of new products. According to Huston and Sakkab (2006, 2007), if a new working system has to be deployed, it is fundamental that it is in line with the company's leadership and with the set of roles, relations, and responsibilities of the individuals and the processes in place. The authors believe that for open innovation to be successful, it is essential that a senior executive is involved. Customer integration in the innovation process may not bring only benefits, but also some disadvantages. These may be avoided by firms if they manage to implement effective strategies, such as those discussed by Enkel, Kausch and Gassmann

(2005). The authors based their research on a questionnaire addressed to 141 firms and investigated in-depth the case studies of nine of them that took part in a number of workshops. Risks and related strategies can be summarized as follows:

- Loss of know-how—reliable customers should be involved, and IP agreements should be implemented at the right time
- Dependence on customers' opinions—the "right" customers should be selected and activity should proceed involving a number of them
- Dependence on customers' demands or personality—exclusivity agreements should be avoided; it is essential to work closely together with HR, and open communication should be used in order to better understand customers' culture and meet their demands
- Limitations to mere incremental innovation—lead and indirect users should be involved, choosing the best way for customer inclusion at the right time
- Serving a niche market only—the strategy should analyze the different phases of the innovation process and focus on the different customers in every stage
- Misunderstandings between customers and employers—customer relations should be developed by utilizing the proper tools and creating an attractive incentive system (figure 1.5).

IP, Patenting, and Appropriation

As stressed by Henkel (2006), the protection of IP is essential for all companies that use open innovation. Chesbrough (2003a) had already stated that IP management by firms is strictly related to the type of paradigm they adopt, that is, whether they operate in a *closed innovation* paradigm or an *open innovation* paradigm. According to the scholar, outside the single firm there is a wealth of ideas, so companies should actively purchase and sell IP. This is the basic assumption of open innovation. The value of the technological assets employed depends on the business model in place. With regard to this, Chesbrough (2003a) presented the cases of Millennium Pharmaceuticals, IBM, and Intel, which utilize diverse strategies to connect IP to business models, and through those models they leverage internal and external IP. Von Hippel and von Krogh (2006) argued that the best way for innovators to make profits from their innovations is free revealing. Innovators could desire to freely reveal information rather than keeping it secret for various reasons: (1) when

Many factors that contribute to make an innovation successful are external to the innovating company (Clark, 1985; von Hippel, 1988; Powell, 1990). Studies on the ways the innovativeness of new products can be enhanced have been particularly lively (Cohen and Levin, 1989). Researchers have revealed how the basic premises of a product technology can be altered by the onset of new technologies, providing novel possibilities to improve performance on a higher level than the one achieved by using classical technologies (Kamien and Schwartz, 1982; Stankiewicz, 1990). At the same time, innovativeness may be indirectly heightened by the number of companies, since, as this grows, also the R&D approaches usually become more differentiated. Moreover, there is an increase in the technology spillover pool and the diversity of technology available for recombination (Ahuja, 1996), and this encourages innovativeness. Several scholars have also begun to analyze the ways the success of new products is influenced by technology exploration and recombination activities of individual companies (see Iansiti and Clark, 1994; Adner and Levinthal, 2001; Katila, 2002; Katila and Ahuja, 2002). Nevertheless, only a small number of longitudinal studies have associated innovativeness of future products with technology search activities. A number of researchers have focused on how innovativeness of new products is increased by utilizing product users. While incentives to engage in product development are influenced by potential market size, market growth (Cooper, 1979a, b), and demand uncertainty (Chaney and Devinney, 1992), here the interest is on how product innovativeness can be improved by resorting to knowledge, feedback, and ideas from users. There are several cases in which users have given birth to ideas that have later turned into new products with a commercial significance (von Hippel, 1988; Zirger and Maidique, 1990; Leonard-Barton, 1995). In other cases, as pointed out by Bresnahan and Greenstein (1996), users have played the role of co-inventors by discovering innovative ways to utilize products or by building something new around product's weaknesses. Several authors have stressed the importance of properly understanding what users need (Cooper and Kleinschmidt, 1986; Zirger and Maidique, 1990; Leonard and Rayport, 1997); others have shown how the lead users of a company can enable its management to anticipate their conclusions with regard to innovative opportunities (von Hippel, 1986; von Hippel, Thomke and Sonnack, 1999). Nonetheless, if it is true that most scholars have identified both technology and user technology as the fundamental drivers of innovativeness promotion, no one has clearly investigated in depth how user knowledge should be searched in order to enhance innovativeness of future products. Moreover, only a few researchers have analyzed how the latter is influenced by the diverse search strategies, with regard to both technology and user dimensions.

Figure 1.5 Technological and user knowledge

something close to their secret is known by other people, (2) when profits from patenting are limited, and (3) when incentives for free revealing are positive. According to the scholars, the presence of free revealing leads one to conclude that there is a private-collective model of innovation incentives that provides the best of both public and private. Henkel (2006) carried out a qualitative and quantitative study of firm-developed innovations with embedded Linux. He discovered that firms realize the importance of protecting their code but also the need to partially reveal it in order to make progress. Thus, according to the amount of external support required to develop their work, they unveil part of the code. Hence, small companies with limited internal resources tend to reveal more. Moreover, in order to minimize competitive loss, companies practice selective revealing. This behavior is in line with profit maximization. According to Hurmelinna, Kyläheiko and Jauhiainen (2007), the decision to be protective or resort to external knowledge is dual. The authors referred to appropriability regime when arguing about the exploitation of knowledge assets and the sustainability of competitive advantage. The regime can be strong or weak, and this may be advantageous or not, as it all relies upon the specific situation the firm is dealing with. The outcomes reveal that legal protection may be the most effective means to solve appropriability issues, and may ensure the firm a higher level of control and different alternatives to exploit arising opportunities.

Industrial Dynamics and Evolutionary Economics

Research in evolutionary economics also suggests that a firm's openness to its external environment can improve its ability to innovate. Studies in evolutionary economics have shown that the ability of a firm to innovate may be enhanced by its openness to the external environment. According to evolutionary economists, search is fundamental for companies in order to discover a number of sources and enable them to re-combine knowledge and technologies in different ways (Nelson and Winter, 1982). As pointed out by Metcalfe (1994), thanks to this variety, companies have the possibility to select diverse technological paths. Moreover, the variety of technological opportunities in the external setting, and the search activities of other companies, have a strong impact on search strategies (Levinthal and March, 1981; Nelson and Winter, 1982). In particular, the environment influences the availability of resources and limits their application. The same effect has the profusion of external knowledge, which can be exploited for innovation purposes. The search strategy of a company may be influenced by both these elements. As stressed by Cohen and Levinthal (1990), the capacity of a company to acquire knowledge from external sources and exploit it widely relies upon the way its internal resources are applied. Even if the intensity of R&D completely depends on internal resources (Greve, 2003), these resources can also have an impact on the potential tradeoffs between exploration and exploitation (Argote and Ingram, 2000). For this reason, the company's search strategies may be affected by limitations on the application of resources.

Another problem associated with open innovation is the availability of external knowledge that is important in the innovation process. This knowledge delimits the space where technological search can be possibly performed. Companies are stimulated to open up their search processes in order to exploit the external sources of knowledge in the industry (Menon and Pfeffer, 2003). In conclusion, the relevant literature on open innovation shows that a company's search strategy is affected by the total availability of external knowledge, the amount of resources that can be exploited for R&D, and the limitations on their application.

In this chapter the concept of open innovation from a firm's perspective has been essentially discussed, while the network or industry levels have been almost disregarded. Vanhaverbeke (2006) stressed the necessity to extend the scope of analysis. Christensen, Olesen and Kjaer (2005) discussed the concept of open innovation in the context of industrial dynamics and applied evolutionary economics. Berkhout et al. (2006) generally highlighted the need of a cyclical model of innovation and stated that nowadays there are four production factors in our society—capital, labor, knowledge, and creativity—on which the "innovation economy" depends. Christensen et al. (2005) claimed that the management of open innovation by firms in relation to innovative technology varies according to (1) their ranking in the innovation system, (2) the phase of maturity of the technological regime, and (3) the value proposition pursued. Many challenges need to be analyzed in-depth in relation to the interaction involving innovative entrepreneurs and incumbents in cases of open innovation when high transaction costs play a significant role. Adopting a regional innovation system perspective, Cooke (2005) stated that companies exploit the regional knowledge capacities in order to bypass intra-firm knowledge asymmetries. This view shows much better than the triple helix model the way research, innovation, and production indeed work. Open science and open innovation are the grounds in which these abilities are rooted. Moreover, skilled knowledge actors gather in "mega-centers," and network nodes are key relay points in global-regional innovation systems.

From a different viewpoint, Bromley (2004) gave account of the great transformation experienced by US manufacturing over the years: craft production, mass production, lean production, and high-quality production. The latest and most impressive change is the one described by Chesbrough (2003b). Nowadays, modern software and the Internet are the means by which companies can make research and get in contact with any party at a global scale able to provide information at the right cost. In this context, Bromley (2004) stressed that for the United States it is essential to thoroughly analyze world trade and technology policy in order to assess how open innovation has influenced its ranking in science and technology and international economic competitiveness.

Chapter Summary

The term "open innovation" is still the subject of much debate, but most scholars believe that the concept is liable to a wider field of application compared to Chesbrough's vision. Today, open innovation can be considered a paradigm that links research from different areas of management sciences. The common goal is understanding how a company can be more innovative, and to reach this aim researchers are expanding in multiple directions and debating on a great variety of fields.

References

Adner, R., & Levinthal, D. Demand heterogeneity and technology evolution: Implications for product and process innovation. *Management Science*, 47(5): 611–628 (2001).

Ahuja, G. Collaboration and innovation: A longitudinal study of interfirm linkages and firm patenting performance in the global advanced materials industry. *Unpublished Doctoral Dissertation*. The University of Michigan (1996).

Argote, L., & Ingram, P. Knowledge transfer: A basis for competitive advantage, in firms *Organizational Behavior and Human Decision Processes*, 82(1): 150–169 (2000).

Ball, A. Open innovation. *R & D Management*, 34(3): 338–339 (2004).

Berkhout, A. J., Hartmann, D., van der Duin, P., & Ortt, R. Innovating the innovation process. *International Journal of Technology Management*, 34(3–4): 390–404 (2006).

Birkinshaw, J., & Gibson, C. Building ambidexterity into an organization. *MIT Sloan Management Review*, 45(4): 47–55 (2004).

Bresnahan, T., & Greenstein, S. Technical progress and co-invention in computing and in the uses of computers. *Brookings Papers on Economic Activity, Special Issue Microeconomics*, 1–77 (1996).

Brierly, P., & Chakrabarti, A. Generic knowledge strategies in the U.S. pharmaceutical industry. *Strategic Management Journal*, 17: 123–135 (1996).

Bromley, D. A. Technology policy. *Technology in Society*, 26(2–3): 455–468 (2004).

Brown, J. S., & Hagel III, J. Creation nets: Getting the most from open innovation. *McKinsey Quarterly*, (2): 40–51 (2006).

Buijs, J. Innovation leaders should be controlled schizophrenics. *Creativity and Innovation Management*, 16(2): 203–210 (2007).

Ebadi, Y. M., & Utterback, J. M . The effects of communication on technological innovation. *Management Science*, 30(5): 572–586 (1984).

Carayannis, E. G., Ziemnowicz, C., & Spillan, J. E. Economics and Joseph Schumpeter's theory of creative destruction: Definition of terms. In Carayannis, E. G., & Ziemnowicz, C. (Eds.) *Re-discovering Schumpeter: Creative Destruction Evolving into "Mode 3"*. London: Palgrave Macmillan: 23–45 (2007).

Chaney, P., & Devinney, T. New product innovations and stock price performance. *Journal of Business Finance & Accounting*, 19(5): 677–695 (1992).

Chesbrough, H. W. The era of open innovation. *MIT Sloan Management Review*, 44(3): 35–41 (2003a).

Chesbrough, H. W. *Open Innovation: The New Imperative for Creating and Profiting from Technology*. Boston, MA: Harvard Business School Press (2003b).

Chesbrough, H. W. The logic of open innovation: Managing intellectual property. *California Management Review*, 45(3): 33–58 (2003c).

Chesbrough, H. W. Managing open innovation. *Research-Technology Management*, 47(1): 23–26 (2004).

Chesbrough, H. W. Open innovation: A new paradigm for understanding industrial innovation. In Chesbrough, H. W., Vanhaverbeke, W., & West J. (Eds.) *Open Innovation: Researching a New Paradigm*: Oxford: Oxford University Press: 1–12 (2006a).

Chesbrough, H. W. New puzzles and new findings. In Chesbrough, H. W., Vanhaverbeke, W., & West, J. (Eds.) *Open Innovation: Researching a New Paradigm*. Oxford: Oxford University Press: 15–33 (2006b).

Chesbrough, H. W. Why companies should have open business models. *MIT Sloan Management Review*, 48(2): 22–28 (2007).

Chesbrough, H., & Schwartz, K. Innovating business models with codevelopment partnerships. *Research-Technology Management*, 50(1): 55–59 (2007).

Chesbrough, H. W., Vanhaverbeke, W., & West, J. (Eds.) *Open Innovation: Researching a New Paradigm*. Oxford: Oxford University Press (2006).

Chiaromonte, F. Open innovation through alliances and partnership: Theory and practice. *International Journal of Technology Management*, 33(2–3): 111–114 (2006).

Christensen, J. F., Olesen, M. H., & Kjaer, J. S. The industrial dynamics of open innovation—Evidence from the transformation of consumer electronics. *Research Policy*, 34(10): 1533–1549 (2005).

Clark, K. The interaction of design hierarchies and market concepts in technological evolution. *Research Policy*, 14: 235–251 (1985).

Cohen, W., & Levin, R. Empirical studies of innovation and market structure. In Schmalensee, R., & Willig, R. D. (Eds.) *Handbook of Industrial Organization*. New York: North-Holland (1989).

Cohen, W. M., & Levinthal, D. A. Absorptive capacity: A new perspective on learning and innovation. *Administrative Science Quarterly*, 35(1):128–152 (1990).

Cooke, P. Regionally asymmetric knowledge capabilities and open innovation exploring 'Globalisation 2'—A new model of industry organisation. *Research Policy*, 34(8): 1128–1149 (2005).

Cooper, R. Identifying industrial new product success: Project new prod. *Industrial Marketing Management*, 43: 93–103 (1979a).

Cooper, R. The dimensions of industrial new product success and failure. *Journal of Marketing*, 8: 124–135 (1979b).

Cooper, R., & Kleinschmidt, E. An investigation into the new product process: Steps, deficiencies, and impact. *Journal of Product Innovation Management*, 3(2): 71–85 (1986).

Dahlander, L., & Wallin, M. W. A man on the inside: Unlocking communities as complementary assets. *Research Policy*, 35(8): 1243–1259 (2006).

Dittrich, K., & Duysters, G. Networking as a means to strategy change: The case of open innovation in mobile telephony. *Journal of Product Innovation Management*, 24(5): 510–521 (2007).

Dodgson, M., Gann, D., & Salter, A. The role of technology in the shift towards open innovation: The case of Procter & Gamble. *R&D Management*, 36(3): 333–346 (2006).

Enkel, E., Kausch, C., & Gassmann, O. Managing the risk of customer integration. *European Management Journal*, 23(2): 203–213 (2005).

Fetterhoff, T. J., & Voelkel, D. Managing open innovation in biotechnology. *Research-Technology Management*, 49(3): 14–18 (2006).

Fleming, L., & Waguespack, D. M. Brokerage, boundary spanning, and leadership in open innovation communities. *Organization Science*, 18(2): 165–184 (2007).

Gambardella, A. *Innovation Inside and Outside the Company: How Markets for Technology Encourage Open Innovation*, p. 85 in Innovation. Perspectives for the 21th century, BBVA Foundation, Madrid (2010).

Gassmann, O., & Reepmeyer, G. Organizing pharmaceutical innovation: From science-based knowledge creators to drug-oriented knowledge brokers. *Creativity and Innovation Management*, 14(3): 233–245 (2005).

Gassmann, O., Sandmeier, P., & Wecht, C. H. Extreme customer innovation in the front-end: Learning from a new software paradigm. *International Journal of Technology Management*, 33(1): 46–66 (2006).

Gaule, A. *Open Innovation in Action: How to be Strategic in the Search for New Sources of Value*. London: Blackwell (2006).

Greve, H. R. A behavioral theory of R&D expenditures and innovations: Evidence from shipbuilding. *Academy of Management Journal*, 46(6): 685–702 (2003).

Gruber, M., & Henkel, J. New ventures based on open innovation—an empirical analysis of start-up firms in embedded Linux. *International Journal of Technology Management*, 33(4): 356–372 (2006).

Henderson, R., & Cockburn, I. Measuring competence? Exploring firm effects in pharmaceutical research. *Strategic Management Journal*, 15: 63–75 (1994).

Henkel, J. Selective revealing in open innovation processes: The case of embedded Linux. *Research Policy*, 35(7): 953–969 (2006).

Hurmelinna, P., Kyläheiko, K., & Jauhiainen, T. The Janus face of the appropriability regime in the protection of innovations: Theoretical reappraisal and empirical analysis. *Technovation*, 27(3): 133–144 (2007).

Huston, L., & Sakkab, N. Connect and develop inside Procter & Gamble's new model for innovation. *Harvard Business Review*, 84(3): 58–67 (2006).

Huston, L., & Sakkab, N. Implementing open innovation. *Research-Technology Management*, 50(2): 21–25 (2007).

Iansiti, M., & Clark, K. B. Integration and dynamic capability: Evidence from product development in automobiles and mainframe computers. *Industrial and Corporate Change*, 3(3): 557–605 (1994).

Jacobides, M. G., & Billinger, S. Designing the boundaries of the firm: From "make, buy, or ally" to the dynamic benefits of vertical architecture. *Organization Science*, 17(2): 249–261 (2006).

Kamien, M., & Schwartz, N. *Market Structure and Innovation*. New York: Cambridge University Press (1982).

Katila, R. New product search over time: Past ideas in their prime? *Academy of Management Journal*, 45: 995–1010 (2002).

Katila, R., & Ahuja, G. Something old, something new: A longitudinal study of search behavior and new product introduction. *Academy of Management Journal*, 45(6): 1183–1194 (2002).

Leonard, D., & Rayport, J. Spark innovation through emphatic design. *Harvard Business Review*, 75(6): 102–114 (1997).

Leonard-Barton, D. *The Wellsprings of Knowledge*. Boston: Harvard Business School Press (1995).

Levinthal, D., & March, J. G. A model of adaptive organizational search. *Journal of Economic Behavior & Organization*, 2(4): 307–333 (1981).

Lichtenthaler, U. The drivers of technology licensing: An industry comparison. *California Management Review*, 49(4): 67 (2007a).

Lichtenthaler, U. Hierarchical strategies and strategic fit in the keep-or-sell decision. *Management Decision*, 45(3): 340–359 (2007b).

Lichtenthaler, U., & Ernst, H. Attitudes to externally organizing knowledge management tasks: A review, reconsideration and extension of the NIH syndrome. *R & D Management*, 36(4): 367–386 (2006).

Menon, T., & Pfeffer, J. Valuing internal vs. External knowledge: Explaining the preference for outsiders. *Management Science*, 49(4): 497–513 (2003).

Metcalfe, J. S. The economics of evolution and the economics of technology policy. *Economic Journal*, 104: 931–944 (1994).

Motzek, R. *Motivation in Open Innovation: An Exploratory Study on User Innovators.* Saarbrücken: VDM Verlag Dr. Müller (2007).

Nagarajan, A., & Mitchell, W. Evolutionary diffusion: Internal and external methods used to acquire encompassing, complementary, and incremental technological changes in the lithotripsy industry *Strategic Management Journal*, 19: 1063–1077 (1998).

Nelson, R. R., & Winter, S. *An Evolutionary Theory of Economic Change.* Harvard University Press: Cambridge, MA (1982).

Nobel, R., & Birkinshaw, J. Innovation in multinational corporations: Control and communication patterns in international R&D operations. *Strategic Management Journal*, 19: 479–496 (1998).

Pavitt, K. Innovating routines in the business firm: What corporate tasks should they be accomplishing? *Industrial and Corporate Change*, 11(1): 117–133 (2002).

Pedersen, T., Soo, C., & Devinney, T. M. The importance of internal and external knowledge sourcing and firm performance: A latent class estimation. In Geisler Asmussen, C., Pedersen, T., Devinney, T. M., & Tihanyi, L. (Eds.), *Dynamics of Globalization: Location-Specific Advantages or Liabilities of Foreignness?*. Bingley: Emerald Group Publishing Limited: 389–423 (2011). *Advances in International Management*, 24.

Piller, F. T., & Walcher, D. Toolkits for idea competitions: A novel method to integrate users in new product development. *R & D Management*, 36(3): 307–318 (2006).

Powell, W. Neither market nor hierarchy: Network forms of organization. In Cummings, L. L., & Staw, B. M. (Eds.) *Research in Organizational Behavior*, Greenwich, CT: JAI Press: 12: 295–336 (1990).

Powell, W. W., Koput, K. W., & Smith-Doerr, L. Interorganizational collaboration and the Locus of innovation: Networks of learning in biotechnology. *Administrative Science Quarterly*, 41: 116–145 (1996).

Rosenkopf, L., & Nerkar, A. Beyond local search: Boundary-spanning, exploration, and impact in the optical disc industry. *Strategic Management Journal*, 22(4): 287–306 (2001).

Rulke, D. L., Zaheer, S., & Anderson, M. H. Sources of managers' knowledge of organizational capabilities. *Organizational Behavior and Human Decision Processes*, 82(1): 134–149 (2000).

Simard, C., & West, J. Knowledge networks and the geographic locus of innovation. In Chesbrough, H. W., Vanhaverbeke, W., & West, J. (Eds.) *Open Innovation: Researching a New Paradigm*: Oxford: Oxford University Press: 220–240 (2006).

Stankiewicz, W. In Sigurdson, J. (Ed.) *Measuring the Dynamics of Technological Change*: 1–25. London: Pinter Publishers (1990).

Surowiecki, J. *The Wisdom of Crowds: Why the Many are Smarter than the Few and How Collective Wisdom Shapes Business, Economies, Societies and Nations.* London: Abacus (2005).

Tao, J., & Magnotta, V. How air products and chemicals "identifies and accelerates". *Research Technology Management*, 49(5): 12–18 (2006).

Tushman, M. L., & Katz, R. External communication and project performance: An investigation into the role of gatekeepers. *Management Science*, 26(11): 1071–1091 (1980).

van der Meer, H. Open innovation—the Dutch treat: Challenges in thinking in business models. *Creativity and Innovation Management*, 16(2): 192–202 (2007).

Vanhaverbeke, W. The inter-organizational context of open innovation. In Chesbrough, H., Vanhaverbeke, W., & West, J. (Eds.) *Open Innovation: Researching a New Paradigm* Oxford: Oxford University Press: 205–219 (2006).

von Hippel, E. Lead users: A source of novel product concepts. *Management Science*, 32: 791–805 (1986).

von Hippel, E. *The Sources of Innovation.* New York: Oxford University Press (1988).

von Hippel, E., Thomke, S., & Sonnack, M. Creating breakthroughs at 3M. *Harvard Business Review*, 77(5): 47–56 (1999).

von Hippel, E., & von Krogh, G. Free revealing and the private collective model for innovation incentives. *R & D Management*, 36(3): 295–306 (2006).

von Krogh, G., & von Hippel, E. The promise of research on open source software. *Management Science*, 52(7): 975–983 (2006).

West, J., & Gallagher, S. Challenges of open innovation: The paradox of firm investment in open-source software. *R & D Management*, 36(3): 319–331 (2006).

West, J., Vanhaverbeke, W., & Chesbrough, H. W. Open innovation: A research agenda. In Chesbrough, H. W., Vanhaverbeke, W., & West, J. (Eds.) *Open Innovation: Researching a New Paradigm* Oxford: Oxford University Press: 285–307 (2006).

Winter Sidney, G. Toward a neo-Schumpeterian theory of the firm, *Industrial and Corporate Change*, 15(1): 125–141 (2006).

Witzeman, S., Slowinski, G., Dirkx, R., Gollob, L., Tao, J., Ward, S., & Miraglia, S. Harnessing external technology for innovation. *Research-Technology Management*, 49(3): 19–27 (2006).

Zirger, B., & Maidique, M. A model of new product development: An empirical test. *Management Science*, 36(7): 867–883 (1990).

Chapter 2

Open versus Closed Innovation: Speculating about the Future of Technology Management

Elias G. Carayannis

Nowadays, in the global economy scenario, technology is fundamental; it covers every aspect of our lives and has substantially altered the way people relate to one another and do business (Del Giudice et al., 2012; Carayannis, 2009). Moreover, the velocity with which technology has developed and spread worldwide is amazing and its rate does not seem to decrease (Chesbrough, 2003; Itami et al., 2010) (figure 2.1).

Nonetheless, if it is true that technology development proceeds at an incredible pace, it is also true that not all countries are able to keep up with it, and this causes the well-known digital divide. Managing to follow the current of technological progress is a challenging issue: economies that tend to stay behind may be helped by institutional development by offering concrete chances to enhance their autonomy and self-sufficiency, also providing them with innovative strategic solutions that can imply better adjustments in the use of technology (Georghiou and Keenan, 2006; Gilsing et al., 2008).

The circulation and use of technology may be intended as a way to make life easier and offer new perspectives. However, it can have positive or negative effects.

When discussing about open innovation, the first company that comes to mind is Procter and Gamble (P&G). As former P&G CEO A. G. Lafley had understood over ten years ago, the company had become almost totally dependent on inorganic growth. Also considering how costly acquisitions were, he realized that the company had to adopt an innovation-driven approach to growth. Internal R&D had always been active, but it was finding increasing difficulties in generating new ideas that could make the difference for a company of such a size. Thus, Lafley believed that external collaboration could be a valid means to speed up the innovation process. To reach this aim, he established the foundation for Connect & Develop (C&D), which is probably the most-well-known corporate open innovation platform at a global scale.

The main objective fulfilled by C&D has been innovation diversity enhancement at P&G. An outstanding example is Olay Regenerist. P&G started a collaboration with Sederma, a French firm leader in the field of active ingredients for the cosmetic industry. Sederma had discovered an amino peptide that accelerates wound healing without leaving traces on the skin. This compound was utilized by P&G to create Olay Regenerist, a product that reduces wrinkles and is now one of the most-well-known treatments worldwide.

Lafley's final objective was to get 50percent of P&G's innovative ideas through C&D. This target was reached over the past ten years. The C&D process is driven by an intricate network, with 70 C&D leaders in different parts of the world, 11 regional hubs, and thousands of sub-networks.

What is the value of innovation for P&G today? As stressed by Bert Grobben, who works at the P&G Singapore Innovation Center, the company cannot do without partnerships. From a historical point of view, P&G has always encouraged broad thinking by hiring individuals from different backgrounds and rotating them across jobs. Thus, as stated by Grobben, P&G has always considered openness to new ideas fundamental, and believes that collaboration is the key to success.

Today, P&G is a $90B company that owns 23 brands that are worth over $1 billion each. The objective is now to create new products that target both luxury and low-spending customers, in order to become even more present in people's lives worldwide. This will definitely challenge C&D since middle class consumers have represented so far the traditional strength of P&G. Current CEO Bob McDonald would like that C&D gave birth to at least 80percent of new ideas.

Nevertheless, as believed by many, P&G's ability to create new high seller products has declined over time, and this suggests that some doubts could arise regarding P&G's dependence on collaboration as an innovation strategy. The basic explanation provided by critics is that the core R&D function within P&G has weakened, and the company is too focused on short-term results. At P&G, different levels of management prevent the company from moving more rapidly in order to compete better with more agile firms. Thus, for P&G to grow, it is necessary not only to take measures in terms of innovation but also to accelerate the decision-making process and general action.

Figure 2.1 Innovation inside and outside the company

At a global level, not all countries have had the same success when attempting to reach the most advanced economies and fill the gap: while some have acquired an increased mastery in using technology, and this ability has caused a real technological revolution in some underdeveloped economies, others have proceeded very slowly and have not been able to catch up.

The process of technology transfer may be more or less effective according to the cultural differences it is defined by, so the mode by which new technology is applied and innovative ideas are produced may be variably successful. For this reason, technology transfer, and its implementation and use, engender more positive outcomes in evolving economies, where they do represent a challenging task.

It is the market that determines the creation of innovative products, and technology industries nowadays are forced to attempt to capture value.

> Even though disruptive technologies initially underperform established ones in serving the mainstream market, they eventually displace the established technologies. In the process, entrant firms that supported the disruptive technology displace incumbent firms that supported the prior technology. The process is understood best by the joint consideration of the trajectories of performance offered by technological alternatives and the trajectories of performance demanded in various market segments.
>
> (Danneels, 2004, p. 2)

Thus, in order to acquire competitive advantage, technology and innovation have become the most important tools, and technology is now the major equalizer among companies and countries (Badawy, 2009). The way the innovation process is managed, and technology is developed and used in business and industry, is fundamental for the success of technology management. However, many hurdles can slow down the pace of conceptualizing, defining and developing the process of converting ideas into real products.

Business is partly stimulated by an organizational culture, which can somehow fuel innovation. The best outcomes leading to an autonomous type of innovation can be achieved with a set of supportive organizational values determined by a change in the organizational culture leading to a reduction of bureaucracy and the development of a more progressive leadership. Innovation can be encouraged by the promotion of a more creative approach, while the reduction of pressures for the provision of products ready to be marketed can work in terms of volumes in the long run (figure 2.2).

In the past, research and development (R&D) activities in China were separated from production and quite restrained in terms of scale, scope, and depth. When China initially adopted the "open door" policy, economic growth was still only partially driven by new knowledge and innovation that played a secondary role and was generally incorporated in the increasing capital stock, which included the early flows of foreign investments.

As pointed out by Gu and Lundvall (2006), the Chinese innovation system was born around the mid-1980s, when the reform of the science and technology system was started up. Relationships between industry and science were fostered by the creation of new infrastructures such as technology business incubators, university science parks, and science and technology industrial parks. Spin-offs from public research organizations were also created. In the 1990s, other factors accelerated the system's maturation: China intensified its opening to the rest of the world (for example, in 2001 it became a member of the World Trade Organization), corporate governance and fundamental premises for innovation were improved (for example, protection of IP rights), and university and public research sectors were further reformed (OECD, 2007).

Starting from the year 2000, a Chinese NIS was being built as a result of a combination of national policies in special zones, a number of initiatives sustained by local authorities, and a set of important systemic reforms. The protection of Intellectual Property Rights (IPR) in China has become a main issue. Because of international pressures and the desire to reach its own economic goals, the Chinese IPR regime has become very similar to those typical of many developed countries. A higher number of IPR infringement cases have been brought to courts; this is the result of the transition of the Chinese economy from manufacturing-based to knowledge-based production, and the enforcement of more comprehensive laws related to IP protection. For the system to work effectively, damage awards related to IP infringements are essential, since they can function as a deterrent to violations and protect the incentives to innovate (Sepetys and Cox, 2009). The Chinese IPR legislation system is quite comprehensive, and generally complies with international standards. In 1982 a Trademark Law was passed, followed by a Patent Law in 1984, and then a Copyright Law in 1990. Despite this legislation, IPR violations occur quite frequently.

Figure 2.2 The Chinese innovation system

The Trends in Evolving Technological Progress as a Field of Inquiry

Technology is never static but it constantly evolves, so keeping up with the new developments is a challenging task.

Technology is not disruptive to society just by its pure existence; it must be embedded into social routines and practices.... we can expect adaptive conflicts. The outcome of these conflicts can lead to the integration of technology into existing social structures or to changes in the social structures that adapt to the technology. Neither way is pre-determined. While many adaptive conflicts are confined to single organizations or to specific industry sectors, some of them become subject to broader public debates and political conflicts. These political interactions play a major role in determining whether the technology transforms society, or is controlled and contains so as to sustain existing social structures and policies.

(Bendrath and Mueller, 2011, p. 1147)

Schools and universities, healthcare facilities, and governments are perpendicular end users of cutting-edge technologies, and they are required to keep pace with technological advancements.

Over the past decades, a huge technological progress has been achieved in a variety of fields, forever changing the world we live in: one need only think of the new tools available on the Internet, such as online media, e-learning, digital marketing, voIP services, and e-banking, just to mention a few examples, but our consideration can also be extended to new technologies in the fields of health care, genetic engineering, biochemistry, nanotechnology, biotechnology, robotics, bio-fuel, artificial brains, aerospace defense industry, high-performance computing, and so on (figure 2.3).

It can be stated without any doubt that, after the steam engine, the invention of the Internet has had the most impact on the history of humankind and has completely changed our world. Centuries seem to have elapsed since the invention of the light bulb, and technological advancement have proceeded at an incredible and unprecedented velocity, permeating in depth our society and altering forever the way we do business. The real novelty compared to the past is the almost instantaneous ability to receive information regardless of place and time: information can reach you wherever you are, at any time, thanks to the surprising opportunities provided by the Internet.

The World Wide Web is a real net of links that allows people and businesses to communicate in real time, increasing the level of global competition. People have become accustomed to the new digital media and have gained experience, so there has been an evolution from simple media literacy to digital media literacy. In this view, one need only think of the impact of e-learning on education, which has radically changed the way students are able to learn. Technology has become widespread and within the reach of everyone, and the Internet has produced major positive consequences considering the possibility of easily accessing any type of information worldwide. The Internet can be deemed as a barrier-free entry to global markets. Information can be transferred, stored, retrieved, and used immediately, since the

In recent times, a number of open source initiatives in the field of medicine have emerged. They have generated innovative and cost-effective information systems on which electronic patient record applications and medical imaging and PACS are based. It has been pointed out that three high barriers have slowed down the adoption of computerized medical records and medical informatics in medicine: excessive cost, the lack of common standards and adequate models, and the transience of vendors. Some scholars, however, believe that these barriers can be reduced by the use of open source software. By this means, in fact, ownership and development costs are cut down and customized tools for clinical practice can be adapted more easily (Caban JJ, Joshi A, Nagy P., 2007). Of course, if the open source software is used, no licensing costs have to be paid. The open source model is welcomed by the medical informatics community, because it is consistent with its model of shared knowledge. Distributors of open source applications can focus on implementation and support services and share development costs with the community of users. From this a real paradigm shift originates: open source medical system vendors do not have to compete with regard to software secrets but can focus on providing a service and compete on that basis. And some specialized areas, such as medical imaging, are even more affected by open source (Erickson B.J., Langer S., Nagy P., 2005). Manufacturers and vendors have always found these markets very engaging: specialized users, in fact, are very demanding with regard to the performance of the tools they require. This has led the market to high-cost developments and marketing strategies that also have to deal with technologies that change quickly and cause rapid obsolescence of most products. Thus, a sustainable return on investment is not always generated (Channin D.S., 2003). In extremely specialized sectors, such as medical imaging, self-sustaining business models from which a profit can be made are almost impossible. In most cases, the costs will be covered through other revenues, such as those from sales of scanners and imaging devices, or by providing costly solutions related to the maintenance of complex integrated information systems (Ratib O., Swiernik M., McCoy J.M., 2003).

Figure 2.3 Open source initiatives

transfer of electronic data is instantaneous. The promotion of cutting-edge technologies and the convenience of their integration are forces that encourage business creation. The entire global economic scenario is constantly evolving due to the rapid progress of the Internet technology.

The potential success of less developed countries that struggle to keep up with the technological progress, and tend to remain behind, can be increased by the use of technological opportunities facilitated by the widespread diffusion of information and the circulation of technology itself.

To fill the gap in terms of technology and innovation between the advanced world and the underdeveloped countries, it is necessary that everyone do their share: challenges can be overcome if the countries lagging behind are assisted in enhancing their abilities in using the new technologies (Mowery et al., 1998; OECD, 2008).

It is paramount that there is a culture willing to receive and adopt the new technologies in order to benefit from them, and end users are the most important beneficiaries. Moreover, to gain competitive advantage it is fundamental that the new technologies are adopted in an early stage.

Opening to receive, assimilate, and implement new technologies with success and trade its products and services is a process of economic transformation. In order to increase business profitability and growth, it is vital to invest on technology management.

Survival and competitive advantage are guaranteed by technology management passing from the production of ideas to their development and implementation, and finally to their commercialization. It is necessary to learn the way to effectively manage technology in order to optimize performance. If a business is willing to survive and be successful nowadays, it has to adapt to changes; in fact, due to all the disruptive technologies, it is of primary importance to reformulate the way of doing business so as not to assume the risk of losing market share.

Technology trends are accelerated, and keeping up with them is a hard task, but many companies do not concentrate enough on the challenges related to technology, and prefer spending their energies on the opportunities arising from it, in order to overcome the digital divide. However, the challenges businesses have to face are numerous and significant, and only by effectively implementing and integrating core technologies it is possible to fill the gap. The greatest successes are achieved by bright leaders who aim at gaining competitive advantage through ad hoc strategies and the continuous and expedited use of a wide range of digital technologies.

Viewing the issue in a prospective way, an integrative risk-taking approach is required in order to determine the path of the process of innovation management in its various shapes. Thanks to the appearance of disruptive technologies, already established technologies are being changed or eliminated, while new ones are being developed. Technological progress and changes constantly occur. A great number of innovation projects consist of a simple adjustment of, re-adaptation of, or small variations in pre-existing technology, while disruptive innovations are determinants of change, and are implemented by technology-driven industries in order to enhance performance and/or gain market share.

However, investing in new technologies has a price, and this is often high, while returns are mainly possible only in the long run, and of course are not certain. Moreover, especially in times of economic downturn, when volatility is the king, the results can be even riskier. Businesses tend to find new possibilities to grow by innovating or acquiring other companies. On the one hand, those that develop new technologies but do not have the assets required to profit from them frequently prefer to commercialize them; on the other hand, others acquire the single inventions or the entire companies, encouraged by reasons of profit, and are also willing to spend a large amount of money to have exclusive rights on the commercialization of the products and services based on specific innovations. They often reduce supplies and increase costs, determining a fictitious scarcity, in order to gain the highest profit possible, creating by this way monopolies and avoiding that the products and services they offer become widespread throughout the market.

From Theory to Practice

According to the definition provided by the *Cambridge Dictionary*, e-commerce can be described as "the business of buying and selling goods and services on the Internet." E-commerce is dynamic and has become a universal trading system. It uses the Internet for the commercial transactions following a linear process that can be personalized. Consumers are constantly directed toward the new marketplace, and businesses are constantly accelerating the use of digital marketing in order to gain new customers. The diffusion of e-commerce is a reality nowadays having moved from the idea of a niche phenomenon in the early years to the creation of an e-world where a great number of people and businesses interact with one another buying and selling. Over a brief period of time it has considerably changed and evolved.

Internet has resulted in a vertical integration of organizational channel capacities such as production, distribution, transaction, and communication and a horizontal integration of organizational communications such as advertising, public relations, and promotion (Li and Leckenby, 2007). Social media channels that emerged from the participatory turn of the Internet facilitated by developments in social computing created new opportunities for interaction and innovation within and across the different stakeholder groups in both the public and the private sector. The increasing adoption and use of social media channels in organizational settings is resulting in a new kind of organizational paradigm termed "social business."

A social business is an organization that strategically engages, analyzes, and manages social media to structure organizational processes and support organizational functions in order to realize operational efficiencies, generate competitive advantages, and create value for customers, shareholders, and other societal stakeholders.

Today consumers have greater power and are able to make the difference, thanks to the innovative tools at their disposal that allow to choose among various options and select the ones that suit them best. The dissemination of information, especially by blogs and online forums, makes companies accountable for their behavior forcing them to be more transparent in their business. Consumers acquire knowledge, become shrewder, and expect more from companies: it is the power of information in all its strength (figure 2.4).

The term "peer production" (PP), also known as mass collaboration, was specifically used within the software industry. Now, instead, its range has extended, and it has become a dominant mode of producing information, which has radically changed how businesses are run. Peer production relies on self-organizing communities of individuals that contribute project components. The common efforts of these people, which are generally Internet based, are coordinated in order to produce a shared outcome. Several types of intellectual output are covered by PP, from software to manuals, books, encyclopedias, reviews, blogs, and so on. Peer production was born within the technology sector, but large companies such as Google and Amazon have discovered its potential and have started to use it. The Internet has made a huge amount of collective outcome available to everyone, and the different projects ranging from user-reviewed databases to constantly edited open source software projects draw from this shared knowledge (The Power of Us, 2011).

Nowadays, especially within the business sector, production is understood in a different way due to PP. One of the basic elements of peer production is the fact that information is widely available, and anyone can contribute to open projects. Open resource initiatives seem to be in contrast with the basic concept of companies that exist to make a profit and do their own research accordingly. As stated by Ronald Coase in 1937, companies exist within markets because they manage to reduce contract costs considerably, as they specialize in a specific field and deal with co-operations all at once. Companies exist only if the market price of their products is higher than the costs of producing them. For instance, Microsoft has an excellent position in its industry because it has "in house" resources gained by hiring its own designers, researchers, and software developers. Thus, Microsoft is able to market its products in a less costly way than the single professionals would have been able to do, if they had individually decided to develop and market the software on their own. Nevertheless, transaction costs are completely erased by peer production. In fact, the hierarchical system based on different management stratification has been made obsolete by the fluid and organic formulations of peer production.

Figure 2.4 Peer production and collective creativity

To reach a more complete understanding of this phenomenon, it is possible to consider one of the most important open source initiatives: Firefox. The basic aim of Mozilla, the company that manages the Web browser, is to "promote openness, innovation and opportunity on the web" (Get to Know Mozilla). The method of production is not determined by any type of hierarchy that dictates the method of production. Any Internet user can give his feedback and take part in the Firefox venture. Today, Firefox is the second most popular Web browser utilized by 20 percent of Internet users (Protalinski, 2012). In June 2008, on its first day of launching, Firefox 3.0 was downloaded eight million times in only 24 hours, and this set a real record. Today Firefox has reached version number 16. Firefox's success is mainly due to its user-friendly interface and the continuous updates and introduction of new features that are a result of the ongoing open source project.

Over time, the value of companies increasingly diminished because of peer production, and also open market had to compete with it. As stated by Yochai Benkler, the Harvard law professor to whom we owe the term *peer production*, "what we are seeing now is the emergence of more effective collective action practices that are decentralized but do not rely on either the price system or a managerial structure for coordination." Thanks to peer production, an ideal market is being created, moving away from the traditional scheme in which a few people tell managers and workers what to do. Decentralization enables information to be exchanged more efficiently through platforms that are based on networking and interdependence. Knowledge is spread among people who now take less time to make better products.

According to Yochai Benkler, individuals who participate in peer production initiatives are not driven by what they can earn from them, but do so "for a wide range of intrinsic and self-interested reasons . . . people who participate in peer production communities love it. They feel passionate about their particular area of expertise and revel in creating something new or better." This seems particularly true in the case of projects such as Wikipedia, an online encyclopedia to which anyone can freely contribute. Nevertheless, since open source is increasingly becoming a significant tool to deal with different types of market competition, firms are becoming more willing to base their strategic infrastructure remodeling on peer production initiatives.

Figure 2.4 (Continued)

If businesses want to survive and be successful, they need to face global competition using all the tools at their disposal. The Internet has revolutionized marketing capabilities, thanks to its interactive nature and the new role consumers acquired within it. This change has completely altered the retail industry, as trade has become more immediate.

Some businesses have achieved very satisfactory results. One of the most famous cases is that of eBay, which was established in 1995, and is now the most important and well-known auction site in the world, with hundreds of millions of users around the globe. Another successful company is Intel Corporation. It already sold over half of its computer chips over the Internet in 1999, and later the majority of its sales was passed onto the Internet.

Similarly, the tourism and hospitality industries have remarkably changed over time and are facing significant processes of marketing evolution, due to the new perspectives inaugurated by the Internet (Mina et al., 2007; Watts and Strogatz, 1998; WRR, 2008; Shih, 2008; Singer and Helferich, 2008).

Retailers are no more hindered by barriers to entry related to geographic distances; the leisure industry has widened its reach thanks to globalization. The traditional agencies functioning as intermediaries have lost a great deal of their business, as their services are not indispensable. The Internet offers all the information needed to directly book a flight to all destinations available, and it is also much easier to compare fares and shop wherever one prefers. Prices, of course, tend to be lower online and this has caused a massive reduction of travel costs.

In the scenario of a very competitive global online market, the time to convert a contact into a customer has become considerably shorter. Consumers have no physical limits on the Web, so they tend to make their decisions much more rapidly and efficiently, as they are able to compare the many options they are surrounded with and choose the best. The Internet provides a great amount of resources to select a location, book a hotel, buy a flight ticket, or rent a car, and all this in a unique online session.

Social Media Networks and Open Innovation: Patterns of Developments

Social media networks are platforms on which people are linked together and interact with one another; nowadays they play an important role in almost everyone's life, and this cannot be merely disregarded or underestimated (Cowan and Jonard, 2004; Kilduff and Tsai, 2003; Kim and Park, 2009; Knoben et al., 2006; Knoke and Kuklinski, 1982; Leenders et al., 2003; Leoncini and Montresor, 2005; Meeus et al., 2008; Midgley et al., 1992).

With all its most utilized and widespread applications, technology has become much less specific than in the past and it is involved in much of our daily routine, enhancing general needs (Carayannis and Campbell, 2009; Carayannis, 1999, 2001). Thus, digital literacy is constantly improving, but also the risks related to the development of digital technologies. In fact, the wide diffusion of online networks has given birth to a new important global reality that needs to be protected from external attacks that could cause damage, so digital tools such as firewalls have to be used in order to avoid similar intrusions.

Life styles and culture are evolving and becoming propelling forces for the creation of new businesses (Leoncini et al., 1996). Emailing, blogging, texting, and podcasting, just to mention some of the new tools, have often taken the place of personal face-to-face interaction. On the World Wide Web, many social media networks have been capable of attracting millions of users in a very short time: among the most surprising examples of such driving forces are Wikipedia, YouTube, Facebook, LinkedIn, Twitter, MySpace, and Yahoo!, only to mention a few.

Facebook, in particular, was created in 2004, and has now become the largest social network worldwide, and Twitter, founded only two years after Facebook, has also been a huge success. Twitter is based on the use of "tweets," short messages with a maximum of 140 characters, which can be posted through website interface, SMS, or a range of applications for mobile devices.

Facebook, Twitter, and also YouTube are not simple tools that enable social networking; they represent the new way people, businesses, and also institutions communicate and interact. All these new media can be very attractive, draw attention, and inform at the same time (figure 2.5).

The relevant literature generally shows that in order to generate innovation it is important that communities collaborate effectively (Winograd, 1996; Williams and O'Reilly, 1998; Lester and Piore, 2004; Baker, 2006, Schneidermann et al., 2006). Evidence is provided by successful design companies, such as IDEO, and products, such as Intuit's Quicken, the Apple Macintosh user interface. SAS, one of the few firms that is constantly creative, stresses the importance of collaboration sustained by effective environments. To do so, it has eliminated differentiations between cultural identities, and has reduced bureaucratic and life distractions as much as possible (Florida and Goodnight, 2005).

As pointed out by Lester and Piore (2004), interpretive conversations among customers, designers, marketers, and other actors are often the grounds from which innovations emerge. In order to stimulate these conversations, firms have taken part in industry consortia and established company-owned stores. Firms utilize environments such as stores to identify high-selling products and focus their production accordingly.

Hence, conversations do not only consist of user feedback regarding specific products, but also direct observation of users' behavior on the spot.

Figure 2.5 Innovation and communities

A study by Von Hippel on user-led innovation reveals the fundamental role played by users in innovation conversations. When lead users are involved, additional value can be created. Lead users not only thoroughly understand the diverse issues that may arise but also have in mind the ways to solve them (Baker, 2006; Franke, von Hippel and Schreier, 2006).

Case studies: Using the social Web to develop new products

DIAMOND CANDLES

Business Challenge: Finding out in real time which candle scents may be preferred by customers in the future.

Project Details: Diamond Candles does not merely utilize traditional market research and trend analysis. The company has found a way to crowdsource submission of new ideas and voting from their customer base. Ten percent of voter suggestions are then taken into account and cross-referenced with market trend analysis in order to finally choose the new candle scents.

Results: Over 250 new product ideas and 5,000 customer votes were submitted in the course of a month. This has set the basis to build an R&D plan.

VITAMIN WATER

Business Challenge: Using Vitamin Water's Facebook fan base to create a new flavor.

Project Details: Vitamin Water's Facebook fan base developed a new flavor called "Connect." A competition was organized and fans took part in the development of many aspects, from flavor selection to package design to the choice of the product's name. One fan, in particular, won a $5,000 prize.

Results: The project involved over two million Vitamin Water Facebook fans.

PRIZE4LIFE

Business Challenge: Identifying a biomarker for amyotrophic lateral sclerosis (ALS), also known as Lou Gehrig's disease or motor neurone disease.

Project Details: Thanks to a collaboration with InnoCentive, an online crowding organization, a $1 million competition was launched by Prize4Life in order to discover an effective way to keep track of the progression of the disease and cut down the costs of clinical trials. About 50 teams from 18 different countries seized the challenge. Dr. Seward Rutkove was awarded the ALS Biomarker Prize by the company's Scientific Advisory Board. Consequently, Dr. Rutkove's research has received a new boost, and many researchers from all over the world have shown their interest.

Results: The cost for a clinical trial was almost halved by the use of Dr. Rutkove's biomarker. Consequently, there has been a definite reduction in the time necessary to ascertain if a particular drug in a clinical trial is beneficial. Moreover, a lower number of patients is required. All this means that possible therapies can be developed more rapidly, the possibilities of discovering a treatment for ALS are enhanced, and it is more likely that resources will be invested in the development of ALS drugs.

Figure 2.5 (Continued)

HOPELAB

Business Challenge: Developing new ideas that can be effective in dealing with childhood obesity through an increase in children's physical activity.

Project Details: HopeLab, a nonprofit organization that deals with the issue of childhood obesity, launched a competition called Ruckus Nation. The aim was to involve people through social networks to provide ideas for products that could increase children's physical activity.

Results: Over 400 ideas were submitted from 37 countries and 41 US states. HopeLab verified that several ideas were likely to be developed to the point that six resulted in patent applications.

MADISON ELECTRIC

Business Challenge: Developing products able to deal with customer issues, satisfy customer demands, or enhance efficiency, on a reduced R&D budget.

Project Details: Madison Electric started up the Sparks Innovation Center, which is the first crowdsourced collaborative approach the company has ever had to product development. Anybody can submit innovative ideas through their Website (www.meproducts.net/sparks).

Every single idea is evaluated by Madison Electric and the best ones are submitted to a focus group through the company's online Contractor Forum.

Results: Almost 100 new ideas have been submitted so far, and five new Madison Electric products have been created, with another four still in the production phase. The Sparks Innovation Center has become a landmark for inventors and aspiring entrepreneurs in the electronics industry.

CDC SOFTWARE

Business Challenge: Resource management across different time zones, countries, and languages to deliver new products and version releases.

Project Details: CDC Software deals with the difficulties in the management of an international development staff spread around the world by developing and delivering software through social network technologies. These are utilized at every single stage of the development process.

Results: Social network technologies have encouraged close collaboration among research and development offices worldwide. Knowledge is transferred efficiently, costs have been reduced, and products are being delivered more rapidly. In fact, delivery time has been cut from 24 months to 12–16 weeks.

CISCO

Business Challenge: Speeding up time to market and time to revenue generation.

Project Details: Cisco created a cross-functional development group called the Cisco Enterprise Collaboration Platform Business Unit (ECP BU). Team members from different functional areas were involved together with their executive sponsors. A collaborative community was created by this team through an internal social platform. By this means, work processes could be easily integrated and product iterations could be rapidly attained.

Figure 2.5 (Continued)

Results: The first and most important outcome was reached by the team within a year. Product time to market was reduced resulting in an average 12 percent increase in productivity per employee, which corresponds to 28,000 labor hours.

GE

Business Challenge: Developing new power grid technologies by leveraging external knowledge and ideas.

Project Details: In 2010, GE launched a global competition called The Ecomagination Challenge, in order to discover new ideas that could be translated into new power grid technologies. GE, thanks to the contribution of important venture capital companies, also invested $200 million in this project. A group of experts evaluated the different ideas, and a panel of judges chose the winners of the $100,000 award, which was given to each creator of the five winning ideas, together with the possibility of collaborating with GE and its venture capital partners.

Results: 70,000 people from over 150 countries took part in The Ecomagination Challenge. A total of 3,844 ideas were submitted and over 120,000 votes were received. Besides the $100,000 award to the five best ideas, other popular ideas were granted a $50,000 cash award, and 12 projects were chosen to partner with GE receiving overall development funds for a sum of $55 million.

Figure 2.5 (Continued)

Discovering Technology Management: Retrospect as a Promising Prospect

The term "technology" dates back to ancient Greece: *technologia* is a Greek word obtained by joining the words *techne*, which means something that belongs to the arts, crafts, or skill, and *logos*, which means logic or science. Consistent with the etymology, technology can be broadly described as the art of logic or the art of scientific discipline. Many definitions of technology have been published (see Steele, 1989; Whipp, 1991; Floyd, 1997). By analyzing the single definitions, it is possible to identify a set of elements that are specific to technology, which can be regarded as a particular kind of knowledge, even if this may be incorporated within a physical device, such as a machine or a product.

From a formal standpoint, Everett M. Rogers defined technology as "a design for instrumental action that reduces the uncertainty in the cause-effect relationships involved in achieving a desired outcome." This definition implies that technology embraces both tangible products and knowledge about processes and methods. For example, it involves both the computer and the technology of mass production implemented by Ford. Michael Bigwood, in his

"Research-Technology Management," highlighted an interesting definition provided by J. Paap. The scholar defined technology as "the use of science-based knowledge to meet a need." According to Bigwood, this definition "perfectly describes the concept of technology as a bridge between science and new products." Technology strongly relies upon scientific progress and findings in research and development. This information is then exploited so the performance and general utility of products and services can be enhanced.

The key feature of technology that differentiates it from more general kinds of knowledge is that it is *applied*, and it specifically focuses on the organization's know-how. Technology itself is mainly related to science and engineering, and in this respect the term "hard" technology is used. However, the processes that enable its effective application also play a fundamental role. With regard to these, the term "soft" technology has been coined: reference is made to new product and innovation processes, along with organizational structures and communication networks. It is useful to consider technology as a type of knowledge, since this allows the application of knowledge management concepts (see Stata, 1989; Nonaka, 1991; Leonard-Barton, 1995; Fleck, 1997; Pelc, 1997; Madhavan and Grover, 1998; Bowonder and Miyake, 2000). For example, technological knowledge normally embraces both explicit and tacit knowledge. Tacit technological knowledge can be described as knowledge that cannot be easily expressed, and widely depends on experience and training. Explicit technological knowledge can be described, instead, as knowledge that has been (or may be) expressed (in the form of a report or a user guide, for instance), along with the physical exhibitions of technology (such as equipment, for example). In the context of firms, technology is viewed as a key resource, so there are valuable connections with other resource-based perspectives of businesses (see Wernerfelt, 1984; Dierickx and Cool, 1989; Penrose, 1995; Grant, 1996), such as competence (Hamel and Prahalad, 1994) and capability approaches (Teece et al., 1997), and the general knowledge management literature.

As for "technology," for the term "technology management" also several definitions have been provided (Roussel et al., 1991; Gaynor, 1996). According to Steele (1989), an integrated view of technology is certainly required. It can be considered a closely connected system that ranges from generating new knowledge to providing post-sales services. It comprises the work of creation and development of products, the processes required for their production or shipment to customers, and overall information processing. In Steele's own words, "technology pervades all aspects of an enterprise, and effective management must recognize its pervasiveness and its crucial role in establishing competitive advantage and even survival."

Starting from the mid-1980s, several commentators have given their opinion regarding the new field of technology management, and many complex definitions have been provided in the literature, but no short definitions have been produced. In fact, it is quite hard to give birth to a short but complete definition of technology management.

If the definitions provided by the most recognized dictionaries are taken into account, technology can be defined, following the *Oxford* dictionary, as "the scientific study of the practical or industrial arts," while according to the *Webster's* dictionary it can be described as: 1—"the branch of knowledge that deals with applied science, engineering, the industrial arts, etc." 2—"The application of knowledge for practical end," while management can be defined as "the act or process of managing." However, all these definitions cannot be deemed complete, since they do not cover all the aspects of technology.

One of the oldest definitions of technology management was suggested by the National Research Council in its 1987 report, which asserted that "management of technology links engineering, science, and management disciplines to address the planning, development, and implementation of technological capabilities to shape and accomplish the strategic and operational objectives of an organization." From that moment onward, several other long definitions have been provided but they are related only to particular subfields of technology management without giving the complete picture, also because at first the latter was not entirely clear. In fact, every definition brought constant changes and an increasing level of sophistication meaning that the field was in continuous evolution.

A quite intriguing definition of technology was proposed by Lowe in 1995. The scholar claimed that

> an ultimate concept of technology is that of socio-technological phenomenon which goes much beyond equipment, labor skills and managerial systems. With such a macro-view, technology includes cultural, social and psychological processes which are related to the central values of a country's culture. The strength of the managerial and social support system is an important factor in the successful international transfer of technology.

As previously stated, technology is dynamic, not static; it is constantly evolving and transforming. Michael Badawy managed to describe more synthetically the process of technology management giving a more explicit definition than others earlier provided in the literature.

He stated: “management of technology is the architecture or configuration of management systems, policies and procedures governing the strategic and operational functioning of the enterprise in order to achieve its goals and objectives.”

Another interesting definition, which can be read in Wikipedia, is supplied again by Michael Badawy, who stated: “IT is concerned with exploring and understanding information technology as a corporate resource that determines both the strategic and operational capabilities of the firm in designing and developing products and services for maximum customer satisfaction, corporate productivity, profitability and competitiveness.” O’Brian, instead, focused on the difference between management information systems and information technology management in the following terms: “Management information systems refer to information management methods tied to automation or support of human decision making.”

A great amount of information relates to the subject of technology management. Most of the definitions found concern single functions and construct isolated compartments, but no definition was discovered that was both brief and all inclusive, considering technology management in its entirety.

After analyzing all the information accumulated over time together with personal considerations and ideas, the only encompassing definition that could be found, responding to the requirement of brevity and comprehensiveness, was the one suggested by A. M. Badawy (2009). The purpose was to provide a comprehensive and short definition of technology management, which was both acceptable and satisfactory. In the author’s imagination, the definition had to match the length limit of 140 characters typical of a tweet, and this operation was certainly possible. The intended limit was exceeded only by two characters, and the ultimate definition of technology management that came up was the following: a “process of effective integration and utilization of innovation, strategic, operational, and commercial mission of an enterprise for gaining competitive advantage.” Of course, the matter is in constant evolution, and the solution proposed will fuel the debate and certainly not prevent others from providing new insights and ideas on the topic.

Chapter Summary

One of the central historical questions concerning technological progress is its extreme variability over time and place. Disruptive technologies have triggered radical changes in system design and IT infrastructures in often unpredictable ways. There is no doubt that the rapid development of the World Wide Web has taken the IT community by surprise and has significantly changed the way information and communications are being handled today. These technologies have also rapidly penetrated more traditional environments and shaped the way traditional applications are being developed today with a massive shift toward Web-based implementations. We tend to be reactive rather than proactive when it comes to studying the problems (and promises) of open innovation that the introduction of new technologies generates. The aim of this section is to provide an innovative definition of technology management. Before reaching this conclusion, it is necessary to position the topic into the correct view and give an account of the role played by technology in the life of individuals, and in a global perspective. The way competitive advantage results from technology and innovation is basically highlighted, presenting some exemplary cases of evolving technological progress, and sequentially focusing on disruptive technologies, e-commerce, tourism and hospitality industries, and social media networks, which have all had a major impact on society. At last, a short definition is provided. In order to perform a significant research, it is necessary to conceptualize the field of technology management to comprehend open innovation.

REFERENCES

Badawy, M. K. A Research architecture for technology management education, A three volume handbook of technology management: Key concepts, financial tools and techniques. *Operations and Innovation Management*, vol. 1, New York: Wiley (2009).

Baker, E. Ideas on the edge, interview: Eric von Hippel. *CIO Insight* (Winter): 12–18 (2006).

Bendrath, R., & Mueller, M. L. The end of the net as we know it? Deep packet inspection and internet governance. *New Media Society*, 1142–1160 (2011).

Bigwood, Michael P. Managing the new technology exploitation process. *Research-Technology Management,* November–December: 38 (2004).

Bowonder, B., & Miyake, T. Technology management: A knowledge ecology perspective, *International Journal of Technology Management*, 19(7/8): 662–684 (2000).

Caban, J. J., Joshi, A., & Nagy, P. Rapid development of medical imaging tools with open-source libraries. *The Journal of Digital Imaging*, 20(Suppl 1): 83–93 (2007).

Carayannis, E., & Campbell, D. "Mode 3" and "quadruple helix": Toward a 21st century fractal innovation ecosystem. *International Journal of Technology Management*, 46(3/4) (2009).

Carayannis, E. G. Fostering synergies between information technology and managerial and organizational cognition: The role of knowledge management. *Technovation*, 19(4): 219–231 (1999).

Carayannis, E. G. The strategic management of technological learning: Learning to learn-how-to-learn in high tech firms and its impact on the strategic management of knowledge. *Innovation and Creativity within and Across Firms*. Boca Raton, FL: CRC Press (2001).

Carayannis, E. G. Firm evolution dynamics: Towards sustainable entrepreneurship and robust competitiveness in the knowledge economy and society. *International Journal of Innovation and Regional Development*, 1(3): 235–254 (2009).

Channin, D. S. Driving market-driven engineering. *Radiology*, 229(2): 311–313 (2003).

Chesbrough, H. *Open Innovation: The New Imperative for Creating and Profiting from Technology*. Boston, MA: Harvard Business School Press (2003).

Cowan, R., & Jonard, N. Network structure and the diffusion of knowledge. *Journal of Economic Dynamics and Control*, 28(8): 1557–1575 (2004).

Danneels, Erwin. Disruptive technology reconsidered: A critique and research agenda. *Journal of Product Innovation Management*, 21(4): 246–258 (2004).

Del Giudice, M., Della Peruta, M. R., & Carayannis, E. *Cross-Cultural Knowledge Management: Fostering Innovation and Collaboration inside the Multicultural Enterprise*, New York: Springer (2012).

Diericks, I., & Cool, K. Asset stock accumulation and the sustainability of competitive advantage. *Management Science*, 35(12): 1504–1511 (1989).

Erickson, B. J., Langer, S., & Nagy, P. The role of open-source software in innovation and standardization in radiology. *Journal of the American College of Radiology*, 2(11): 927–931 (2005).

Fleck, J. Contingent knowledge and technology development. *Technology Analysis & Strategic Management*, 9(4): 383–397 (1997).

Florida, R., & Goodnight, J. Managing for creativity. *Harvard Business Review* (July–August) (2005).

Floyd, C. *Managing Technology for Corporate Success*. Aldershot: Gower (1997).

Franke, N., von Hippel, E., & Schreier, M. Finding commercially attractive user innovations: A test of lead-user theory. *Journal of Product Innovation Management*, 23: 301–315 (2006).

Gaynor, G. H. (Ed.) *Handbook of Technology Management*. New York: McGraw-Hill (1996).

Georghiou, L., & Keenan, M. Evaluation of national foresight activities: Assessing rationale, process and impact. *Technological Forecasting & Social Change*, 73: 761–777 (2006).

Gilsing, V., Nooteboom, B., Vanhaverbeke, W., Duysters, G., & van den Oord, A. Network embeddedness and the exploration of novel technologies: Technological distance, betweenness centrality and density. *Research Policy*, 37(10): 1717–1731 (2008).

Grant, R. M. Toward a knowledge-based theory of the firm. *Strategic Management Journal*, 17: 109–122 (1996).

GU, S., & Lundvall, B-Å. Policy learning as a key process in the transformation of the Chinese Innovation Systems. In Lundvall, B-Å., Intarakumnerd, P., & Vang, J. (Eds.) *Asian Innovation Systems in Transition*. Edward Elgar Publishing Ltd: (2006).

Hamel, G., & Prahalad, C. K. *Competing for the Future*. Boston: Harvard Business School Press, (1994).

Itami, H., Kusunoki, K., & Numagami, T., et al. (Eds.). *Dynamics of Knowledge, Corporate Systems and Innovation*. Springer (2010).

Kilduff, M., & Tsai, W. *Social Networks and Organisations*. Thousand Oaks CA: Sage (2003).

Kim, M. S., & Park, Y. The changing pattern of industrial technology linkage structure of Korea: Did the ICT industry play a role in the 1980s and 1990s? *Technological Forecasting and Social Change*, 76(5): 688–699 (2009).

Knoben, J., Oerlemans, L. A. G., & Rutten, R. P. J. H. Radical changes in inter-organizational network structures: The longitudinal gap. *Technological Forecasting and Social Change*, 73(4): 390–404 (2006).

Knoke, D., & Kuklinski, J. *Network Analysis*. London: Sage (1982).

Leenders, R. T., Van Engelen, J. M. L., & Kratzer, J. Virtuality, communication, and new product team creativity: A social network perspective. *Journal of Engineering and Technology Management*, 20(1–2): 69–92 (2003).

Leonard-Barton, D. *Wellsprings of Knowledge—Building and Sustaining the Sources of Innovation*, Boston: Harvard Business School Press (1995).

Leoncini, R., & Montresor, S. Accounting for core and extra-core relationships in technological systems: A methodological proposal. *Research Policy*, 34(1): 83–100 (2005).

Leoncini, R., Maggioni, M. A., & Montresor, S. Intersectoral innovation flows and national technological systems: Network analysis for comparing Italy and Germany. *Research Policy*, 25(3): 415–430 (1996).

Lester, R. K., & Piore M. J. *Innovation: The Missing Dimension*. Cambridge, MA: Harvard University Press (2004).

Li, H., & Leckenby, J. D. Examining the effectiveness of internet advertising formats. In Schumann, D. W., & Thorson, E. (Ed.) *Internet Advertising—Theory and Research* (First edition, 203–224). Mahwah, NJ: Lawrence Erlbaum Associates (2007).

Lowe, P. *The Management of Technology: Perception and Opportunities*. United Kingdom: Chapman & Hall (1995).

Madhavan, R., & Grover, R. From embedded knowledge to embodied knowledge: New product development as knowledge management. *Journal of Marketing*, 62: 1–12 (1998).

Meeus, M. T. H., Oerlemans, L. A. G., & Kenis, P. Inter-organisational networks and innovation, In Nooteboom, B., & Stam, E. (Eds.) *Micro-Foundations for Innovation Policy*, Amsterdam: University Press (2008).

Midgley, D. F., Morrison, P. D., & Roberts, J. H. The effect of network structure in industrial diffusion processes. *Research Policy*, 21(6): 533–552 (1992).

Mina, A., Ramlogan, R., Tampubolon, G., & Metcalfe, J. S Mapping evolutionary trajectories: Applications to the growth and transformation of medical knowledge. *Research Policy*, 36(5): 789–806 (2007).

Mowery, D. C., Oxley, J. E., & Silverman, B. S Technological overlap and interfirm cooperation: Implications for the resource-based view of the firm. *Research Policy*, 27(5): 507–523 (1998).

Nonaka, I. The knowledge-creating company. *Harvard Business Review*, November–December: 96–104 (1991).

OECD. *Policy Brief: Open Innovation in Global Networks*. Paris: OECD (2008).

Pelc, K. I. Patterns of knowledge generating networks. *Proceedings of the Portland International Conference on Management of Engineering and Technology (PICMET)*, Portland, July 27–31 (1997).

Penrose, E. *The Theory of the Growth of the Firm*. Oxford: Oxford University Press.

Ratib, O., Swiernik, M., & McCoy, J. M. From PACS to integrated EMR. *Computerized Medical Imaging and Graphics*, 27(2–3): 207–215 (1995).

Rogers., & Everett, M. *The Diffusion of Innovations*. New York: The Free Press (1995).

Roussel, P. A., Saad, K. N., & Erickson, T. J. *Third Generation R&D—Managing the Link to Corporate Strategy*. Boston: Harvard Business School Press (1991).

Schneidermann, B., Fischer, G., Czerwinski, M., et al. Creativity support tools: Report from a US National Science Foundation sponsored workshop. *Journal of Human-Computer Interaction*, 20(2): 61–77 (2006).

Sepetys, K., & Cox, A. *China: Intellectual Property Rights Protection in China: Trends in Litigation and Economic Damages*. NERA Economic Consulting (2009).

Shih, H. Y. Contagion effects of electronic commerce diffusion: Perspective from network analysis of industrial structure. *Technological Forecasting and Social Change*, 75(1): 78–90 (2008).

Singer, J., & Helferich, J. Social network analysis can help managers build better connections between support groups and their customers. *Research Technology Management*, 51(1): 49–57 (2008).

Stata, R. Organizational learning—the key to management innovation, *Sloan Management Review*, Spring (1989).

Steele, L. W. *Managing Technology: The Strategic View*. New York: McGraw-Hill (1989).

Teece, G., Pisano, G., & Shuen, R. Dynamic capabilities and strategic management. *Strategic Management Journal*, 18: 509–533 (1997).

Watts, D. J., & Strogatz, S. H. Collective dynamics of small-world networks. *Nature*, 393(6684): 409–410 (1998).
Wernerfelt, B. A resource-based view of the firm. *Strategic Management Journal*, 5: 171–180 (1984).
Whipp, R. Managing technological changes: Opportunities and pitfalls. *International Journal of Vehicle Design*, 12(5/6): 469–477 (1991).
Williams, K. Y., & O'Reilly, C. A. Demography and diversity in organizations: A review of 40 years of research. *Research in Organizational Behavior*, 20: 77–140 (1998).
Winograd, T. Introduction. *In Bringing Design to Software*. New York: ACM Press (1996).
WRR. *Innovatie vernieuwd: Opening in viervoud*. Amsterdam: Amsterdam University Press, (2008).

Chapter 3

Invention, Inventiveness, and Open Innovation

Manlio Del Giudice

This chapter will provide a list of definitions in order to better illustrate the terminology used in the chapter title.

Schumpeter (1934) first emphasized the dual significance of invention and innovation. His seminal early works of the 1930s and 1940s had described entrepreneurship as an act of "creative destruction" (of market equilibrium), which carried an invention or creation into innovation, resulting in a new business that could grow successfully (Kirchoff, 1991, p. 8).

> Inventions are very commonly the result of combining or recombining existing elements of knowledge into new syntheses.
>
> (Ahuja and Lampert, 2001, p. 528)

According to Bozeman and Link (1983, p. 4), "The concepts commonly used in connection with innovation are deceptively simple. Invention is the creation of something new. An invention becomes an innovation when it is put in use." Innovations may be new products, new processes, or new organizational methods that are novel and add value to economic activity. Thus, at a general level, invention parallels the concept of science and innovation parallels the concept of technology.

As Scherer and Ross (1990) observed, "Technical innovations do not fall like manna from heaven. They require effort—the creative labor of invention, development, testing and introduction into the stream of economic life." While most innovations may be serendipitous, firms may have a comparative advantage in generating inventions because they "are more likely to put together the critical combination of a fertile mind, a challenging problem, and the will to solve it" (Scherer and Ross, 1990).

Invention, and in particular technological invention, is the process of conceiving and producing, through autonomous study and experimentation, a device or method that is new and useful.

Inventiveness is the ingenious vision that leads to invention. Even if invention, in its broadest sense, embraces more than technological invention, as for example the invention of organizational structures, this study focuses specifically on technological invention and inventiveness. The outcomes of technological invention can be various and diverse, not only devices and machines, but also methods and processes, databases and algorithms. Design is based on invention, while at the other end of the scale is routine problem solving. Routine problem solving is related to the increase in specificity and predictability, while invention is related to the increase in boundary transgression.

Boundary transgression is associated with mental patterns that go beyond conventional paths and practice, so disciplines are bound together in an unusual fashion, well-rooted beliefs are challenged, and issues are redefined in innovative ways. Macro-inventions are inventions of such significance that they alter the ordinary way of living. Several improvement inventions, called micro-inventions, ensue from them. Although many of these micro-inventions may never be patented and commonly utilized, they are true examples of inventiveness and creativity.

Innovation is the complex act of starting something new and implementing it practically. Entrepreneurship is to be considered fully part of this process. Normally invention is deemed significant only if its outcome is widely used. Hence, benefits to society originate not from invention alone, but particularly from innovation. Much research has focused on the importance of innovation to society. However, this study takes specifically into account invention, which can be described as the source of innovation.

Human History and Technological Progress

Inventiveness is inherent in the human species, but in the distant past, invention was rather occasional and certainly not widespread or persistent.

Inventions started to become more common and sustainable after the scientific revolution (approximately from 1520 to 1750) and the first Industrial Revolution (approximately from 1760 to 1850). Technology was the main determinant of the first Industrial Revolution, which began in the mid-eighteenth century. There was a higher level of access to knowledge based on discovery and invention, and a stronger foundation for additional inventions was built by the feedback process between discovery-based knowledge and invention-based knowledge. Nevertheless, the discovery-based knowledge of the time was widely pragmatic, informal, and empirical. The second Industrial Revolution, which started after the Civil War and witnessed the birth of corporate research laboratories, was a period in which the quantity of inventions increased dramatically, as proven by the number of patents issued. This was mainly due to applied science, which had considerably improved in the nineteenth century (Dennis, 2004; Dickson and Noble, 1981; Dickson, 1988).

The best way to understand this process is to relate it to the new scientific vision of discovery-based knowledge on which invention was based; unlike what had occurred in the past, it has now become more formal and consensual

Inventions originate from the continuous interaction between technical or cultural requirements and individual or institutional will. Inventions are created by people who use their imagination that interacts with the most significant values and issues in their daily lives. Inventions cannot work alone; they need complementary technologies, so it is important to consider the systems environment in which both invention and innovation occur (Drori, Meyer, Ramirez and Schofer, 2003; Economist Intelligence Unit, 2004).

Inventions are commonly distinguished into *macro-inventions* and *micro-inventions. Macro-inventions* are responsible for important changes in society, go beyond their initial sphere of application, and encourage a great number of micro-inventions. *Micro-inventions* involve the changes in the process and product on which research and development (R&D) is mainly based (Graham, 1985). Through micro-inventions an originally raw idea can be finally considered commercially viable, and its application may be extended to areas that the inventor had not initially foreseen. Both micro- and macro-inventions are strictly linked in an interactive system in which they enable each other. It is important to stress that the distinction can be made only in retrospect; however, it is essential to acknowledge that the scale and scope of inventions may vary. The path taken by inventiveness in society depends on the role played by economic forces and the way these decide to support research and development. Governments have often encouraged inventiveness by financing projects in which R&D managers have been able to allow ample space for imagination and interaction. Discovery-based knowledge has been extended thanks to governmental support of individual researchers; however, new ways to improve invention have not been so easily unveiled.

The role of the inventor in public research demands attention because "[t]here also exists the possibility that returns to individual inventors could influence decisions at the bench scientist's level. Many public institutions grant inventors a portion of the licensing proceeds. Thus individual scientists could be motivated by income from patenting and licensing, even if the employing institution is not" (Rubenstein, 2003, p. 115).

> [O]n a micro level scientists decide whether to publish their results and place them in the public domain, or to begin the process of seeking intellectual property. If a scientist makes a major discovery, the inventor's percentage of ARS licencing fees and royalties could motivate the scientist to opt for patenting over publishing.

And the purpose of the "Patent Review Committee is to determine whether patenting is the most beneficial route for disseminating an innovation" (Rubenstein, 2003, p. 127).

In the past, academic contexts have been stimulated by public policies to become a forge of inventions. Also, inventiveness has been encouraged through public education funding at different levels. Among the first examples of similar measures, it is possible to recall the Morrill Land Grant Act of 1862 according to which American colleges and universities were granted land, and the creation of the public school system in the USA. Another example is the Servicemen's Readjustment Act of 1944, known as G. I. Bill, that granted considerable education benefits to the war veterans and their families. They both paved the way for the American economic growth over the years, contributing to make the USA the most powerful country in the world.

Nevertheless, these interesting topics and their implications for invention, together with their impact on the US economic growth, have not been the subject of thorough systematic research. The most genial engineers and scientists are generally assisted by researchers to whom significant contributions to the process of discovery and invention can be traced back. These people are an extraordinary resource for their country, and much has been done by the American government to guarantee the existence and prosperity of a similar infrastructure.

Inventiveness and creativity in the fields of science and engineering have been stimulated by the creation of flexible learning contexts. History has revealed several examples of individuals who became well-known inventors and scientists, and were encouraged and supported, both at home and in school, by different people, from parents to teachers, who were willing to do so only because they were truly interested and committed. However, no systematic research has ever thoroughly investigated these dynamics.

In the past, women, blacks, and other minorities were basically denied participation in the invention process due to cultural biases and segregation policies. This preclusion has been overcome only in comparatively recent times, and the utmost has to be done to guarantee that anyone who has the right credentials can start a career in science and invention. A modern society requires openness but above all tolerance. Invention can be often viewed as a way an artist or a creative engineer does not conform to the established system and rebel against it.

Much scientific discovery and engineering invention was born in the USA. Nevertheless, the wider and long-lasting consequences of the adoption of new technologies have not been always adequately foreseen. New technologies usually imply issues that should be anticipated and dealt with proactively; instead, the common behavior is to react to problems only when they arise. The institutional nature of invention itself has considerably evolved over time. For instance, the world today is one in which physics and engineering play a key role, and the impact of this transformation for the government, academic institutions, businesses, and society in general is tremendous.

TECHNOLOGICAL INVENTION VERSUS SCIENTIFIC INVESTIGATION

Mokyr (1990, p. 170) claimed that in the past 150 years, a majority of inventions were used before people understood why they worked. This occurred in agricultural technology, mechanical machinery, metallurgy, the textile industry, and shipping. However, increasingly, from the second half of the nineteenth century, scientific understanding came to feed technological development. An example is the invention of telegraph, which required the theoretical notion of electromagnetic waves invented by Maxwell in 1865 (Mokyr, 1990, p. 144). Chemistry was also to a large extent guided by science. Nevertheless, the reverse order of practical tinkering preceding understanding still occurs, as has been exhibited in the information technology revolution (Dosi, 1984).

Technological invention differs from scientific investigation because it focuses on the development of products or methods that have a practical use. Scientific theories and findings, instead, elaborate clean models that exclude complications in their analyses.

The spectrum of invention is wide. Technological invention may result in new physical devices, machines, processes, databases, algorithms, and so on. Their impact on society can be different. In some cases, social impact is strong and, as a result, society is completely changed, as it happened in the case of automobile. With regard to knowledge extension, some inventions are based on acquired knowledge, while others, as in the case of nanotechnology, need thorough investigation. On a systemic level, some appear at the level of components, others at the product or process architecture level, and others involve entire systems.

It is interesting to study how an inventive mind expresses itself and how inventive work gets done by groups of individuals in society. From an historical point of view, most inventions cannot be traced back to their original inventor. They just circulated successfully over time, while they were subjected to modifications and improvements; thanks to several contributions from people, it is no longer possible to track back. The development of modern agriculture is one of the best examples that illustrates the idea. Human existence is largely based on constant inventiveness, but only in recent centuries there has been a clear social definition of the role of inventors. Nowadays, inventors have an outstanding position in society, and the invention process and context can be easily investigated. The social, economic, cultural, and institutional environments together are the contexts in which invention occurs. Inventors have to deal with two fronts, nature and society. First, they have to understand the materials and natural processes deployed. Then, they have to create inventions that have a practical use and are commercially viable. In some societies more than in others, inventiveness has been intensely stimulated, and this shows the values and needs of those societies. At times, inventors attempt to meet social needs by dealing with issues that have already been acknowledged; however, in other cases, they break down a problem and invent a new solution that could lead to a new opportunity. From an economic point of view, it could be stated that in the first instance, inventions are pulled by demand, in the sense that inventions are a response to demands already present on the market. In the second case, inventions are pushed by supply, in the sense that they are generated from the solutions devised by the inventor, with the result that a need is created, and consequently people have to be convinced they have that need and must use the invention to fulfill it.

Successful inventors are characterized by initiative, resilience, and noncompliance; they are practical, enthusiastic about their work, optimist, and patient, and they tolerate complexity and do not mind delaying gratification. Effective inventors have a critical attitude toward their own work. Failure is to be considered an experience to learn from. Knowledge that is deemed too restrictive is easily abandoned. They take into account practical problems, tackling them by using a number of measures to break them down from a conceptual viewpoint and find effective solutions. Typically, inventors have profound knowledge in their areas of expertise on both a theoretical and a practical basis; at the same time, bypassing the boundaries of established knowledge is not a problem for them. This kind of attitude usually triggers high performance and is responsible for the most creative endeavors. Responsiveness to practical issues and opportunities, together with the theoretical and functional knowledge, is fundamental for technological invention. It is essential to highlight the predisposition of the inventive mind, the way it is curious, enthusiast, recognizes issues and opportunities, and is committed to find solutions. Although knowledge and abilities variously combined are considered fundamental with regard to the inventive mind, it should be stressed that they are not sufficient to explain inventiveness. In fact, for the progress of their inventions, a set of other skills related to the external environment is required by inventors. Of course, inventors mainly focus their attention on the invention itself, but other skills are often involved, associated with persuasion and promotion, since financial resources have to be raked up and the invention has to be also marketed. In some contexts, these roles may be assigned to other people, but in others the responsibility for those tasks is taken by the inventors themselves; for example, in order to make their endeavor progress within an organization, they often have to act as "intrapreneurs."

Inventors are sometimes seen as less educated than technical experts, but research shows that successful inventors, regardless of their education, have always a deep knowledge of their areas of expertise both theoretically and practically. They count on knowledge extracted from diverse disciplines, in relation to what is required for the advancement of their project, and sometimes they go beyond the boundaries of what is well established. Research has proven that for an individual to work at a real expert level, at least ten years of experience in a specific area of expertise are required to gain the quantity and quality of knowledge related to it. Successful inventors are not constrained by the current level of knowledge; they are boundary transgressors. They activate their knowledge in a flexible and critical way. They select what is already known and manage to exclude approaches that are not deemed useful while questioning those that could be false or defective. Inventors typically rely upon a combination of profound theoretical understanding of materials and processes and practical knowledge of how things work in the physical and social environments. The former is characteristically systematic and complex; the latter is often definitely founded on experience and difficult to reveal by the use of simple formulas. Differences in balancing the two may affect particularly inventive endeavors.

The way an invention may develop widely relies upon the possibility of accessing adequate knowledge resources (Guena, Salter and Steinmueller, 2003).

These can consist of journals, reports, technical manuals, patent descriptions, the Internet, the availability of samples and prototypes, and the wisdom of other experts and collaborators. A prompt and adequate flow of knowledge is essential to the invention process. In many respects, the commonalities between technological invention and other creative endeavors are more surprising than the differences. High commitment, effort, and persistence are features common to all highly performing firms, whether they are creative or not. Other common characteristics are the trend toward autonomy and flexibility, and a number of boundary transgressions together with a variety of ordinary problem-finding and problem-solving issues.

The commitment to create something that can be used in practice intensely shapes the inventive mind, so inventors have to face a number of considerations associated with the need to make something actually function in the physical and social world. This means that an invention not only needs to work physically, but also needs to be economically viable, function within fair time and space boundaries, and not cause risks to its users. Scalability is usually considered important in this respect.

The aim of a scientist is to understand, and his or her actions are based on the idea that an explanation always exists. A scientist's duty is to comprehend what nature is doing. An inventor attempts to find a solution although he or she does not know a priori if it exists. It is possible that the inventor has no knowledge of the problem (Feldman, Link and Siegel, 2002; Foray, 2004). Maybe the job cannot be done within reasonable cost and time parameters or cannot be done at all. Therefore, differently from the life of a scientist, the life of an inventor is uncertain. And an inventor is also different from an artist who is able to create something that is always viable, regardless of its quality. An invention does not function the same way as a work of art. Invention, which is the true expression of a creative mind, is also a powerful motor of social progress.

EDUCATION AND PROCESS OF INVENTION

Education together with proper societal support may enhance the features of the inventive mind and improve the process of invention. However, the educational system could also undermine these potential results, as it is so common nowadays. The fundamental question concerns the role education will play in the near future: will education primarily foster or hinder inventiveness?

The latter may be cultivated in different contexts and at various levels. In formal education, all students should have the possibility of better understanding how invention works, and gain basic knowledge and skills. Those who prove to be especially inclined toward invention deserve to take advantage of the opportunity to learn more and make further progress. Nevertheless, inventiveness not only may be developed in a formal education setting, but also may be stimulated or discouraged by institutional cultures and patterns of practice in any type of environment, from classrooms to corporations. Hence, the following question arises: how can the various tensions and doubts regarding the encouragement of inventiveness be dealt with by schools, universities, and informal educational contexts? The inventive mind may not be adequately stimulated by traditional teaching with its pre-established items and principles. In parallel with the classical approach, other educational means may be similarly significant, such as mentoring, modeling, project-based learning, and so on. The agenda is partially made up of the global structure of inventive activity, that is, its enduring life span, its usefulness in a flexible manner, and its ability to find problems in addition to solving them. From a dispositional point of view, inventiveness requires curiosity and confidence in the possibility of exploiting new opportunities together with desire to take risks. At the same time, it is important to focus on the need to avoid repressing failure, dejecting challenge, and organizing learning experiences in a mechanical way. Although this is true with regard to fostering any kind of creativity, the typically inventive side of invention must not be neglected. This involves interaction between abstract thought and practical investigation, and highlights the importance not only of scientific knowledge but also of operational principles together with distinct levels of inventive thought.

An important input to invention is scientific knowledge (Fuller, 2000a, 2000b, 2002; Galison and Hevly, 1992; Gaudilliere and Lowy, 1998; Geiger, 1997). Scientific knowledge may be gained by collaborating with university scientists or reading academic publications. There is considerable evidence that industrial breakthroughs are related to both knowledge in the public domain and participation in scientific research (Darby and Zucker, 2003; Henderson and Cockburn, 1994; Narin, Hamilton and Olivastro, 1997; Thursby and Thursby, 2006; Zucker, Darby and Armstrong, 2002). Profound technical knowledge is needed nowadays to create inventions in several fields, and it is likely to be provided by modern universities. Nevertheless, this does not suffice, since inventiveness also depends on creativity capability. Universities can less easily encourage this important quality in young people. In academic contexts, deductive learning is too often overstated, principles are learnt regardless of their application, self-discovery is not adequate, formats are too rigid, results are preset, learning from failure is not given the space it should, and teaching through open-ended problems and visual thinking are lacking. In a few instances, many of these issues have been bypassed with success. This is the case of educational inventions, such as community invention centers or educational courses that are tremendous experiences that change students' lives. However, these innovations have not been adequately supported, so they are often one-time experiences that have not become widespread. The limited circulation of these innovations is partly due to the lack of adequate mechanisms that could assist new instructors in developing the ability to stimulate inventiveness. At the same time, mechanisms through which instructors who are innovating in this area may be brought together are missing. Behind all this is the fact that invention and teaching of inventiveness are only seldom rewarded by the academic incentive system. On the contrary, these activities are often discouraged in a direct or indirect manner. The education system has to face inherent tensions when approaching the matter of inventiveness encouragement. Thus, it is possible to list a series of dilemmas: the weight of individual versus group effort in invention, the need for reflection versus rapid investigation, the fundamental roles of cooperation versus competition, the importance of disciplinary expertise versus open-ended analysis, and the roles of intrinsic versus extrinsic motivation. The tensions incidental to every dilemma must be dealt with effectively by the education system, and this implies difficult decisions in order to make inventiveness advance.

Although inventors are commonly seen as solitary actors, the process of invention is profoundly collaborative (Gibbons, 2003;

Gibbons, Limoges, Nowotny, Schwartzman, Scott and Trow, 1994). The universities are not only places where advanced research and education are integrated on a large scale, but places where a high number of diverse ideas and different interdisciplinary approaches come to light. The academic research endeavor is constantly refueled by the arrival of new students that bring their new ideas and question the pre-established setting. Additional intellectual challenges are determined by collaborations with businesses and industrially supported research. These create an environment that may encourage inventiveness in perspective, since it unites problem formulation, boundary transgression, focused effort, and open, creative minds.

Both universities and businesses have benefited from many advantages associated with the higher attention paid by universities to invention. Among these, it is possible to mention the provision of incentives for inventions by students, the technique of teaching invention "by doing," and the promotion of the commercialization of new technologies. Of course, many tensions arise around issues such as the weakening of the wider academic mission constrained by rigid definitions of excellence, financial problems, and inherent uneasiness related to the assessment of faculty effort. Auerswald and Branscomb (2003) state that "the process by which a technical idea of possible commercial value is converted into one or more commercially successful products—the transition from invention to innovation—is highly complex, poorly understood, and little studied." They provide a model and point to places where gaps exist in the support needed for commercialization and indicate the "capital gap as a major obstacle for new technology development" between new technologies and commercial application (Auerswald and Branscomb, 2003, p. 229).

This is a fundamental time for global support of inventiveness by society in general, and the educational institutions in particular (Etzkowitz, 2002, 2003; Etzkowitz and Leydesdorff, 2000; Etzkowitz, Webster and Healey, 1998; Etzkowitz, Webster, Gebhardt and Terra, 2000).

Ownership Rights for Inventions

Invention is much older than the concept of intellectual property (IP). New tools, techniques, and technologies had been produced ages before individuals and organizations were legally granted ownership rights for their inventions. Patenting was understood as a mechanism of reward and motivation, established also to publicize new ideas and contribute to the general progress of society. Patenting systems have considerably evolved over time. Historically, among the most relevant legal measures in this respect, it is important to recall the first patent law that dates back to the fifteenth-century Venice, not to mention the seventeenth-century English patent statute and the US system of patent protection established in the last decade of the eighteenth century. Today the international patent structures have changed, together with the need to ensure protection of additional forms of IP, such as copyrights and trade secrets.

This study attempts to respond to general questions such as: is the creative process of invention effectively sustained by the current system of intellectual property? How can it be enhanced successfully? Customer of the patent system is the entire society. Societal economic welfare is largely dependent on high levels of invention, and invention is sustained by the patent system. Patents reward inventors, since these are enabled to economically benefit from their inventions after externalizing their know-how to society. This has positive effects: investment in new technologies is encouraged, companies are stimulated to invent new products, and entrepreneurs are motivated to start up new businesses. Although inventors are primarily encouraged by the potential financial revenues ensuing from the patent, other motivations may be taken into account, such as selflessness, the simple delight of inventing, and professional acknowledgment.

In the past two decades, a series of legislative acts and judicial decisions have widened and reinforced patent rights in the USA. Patenting is evolving as it has to deal with new fields such as gene manipulation, new types of software, or innovative business methods, just to mention a few. One industrial sector in particular accounts for most of the recent expansion of patent rights: electronics, computing, and communications. Most of the patents are filed for defensive reasons, since big players require adequate protection of their IP or are willing to make a profit from trading their portfolios. The number of software patents has increased dramatically over the past two decades. Software inventions may be also protected by trade secrets, and the source code may be protected by copyright.

The importance and value of patents vary considerably. Less than 10 percent are commercially relevant, and less than 1 percent are of pivotal importance. It is assumed that the most mentioned patents are also the most valuable. They may be referred to in papers or other patent applications. Between 1975 and 1998, 85 percent of the highly referenced patents were granted to corporations, 9 percent to individuals, 4 percent to universities, and 2 percent to government and nonprofit organizations. In fact, 0.5 percent of patents granted in the period 1963–1999 that are quoted more than five standard deviations above the average for patents assigned those years were granted to US corporations; that is almost 70 percent compared to 46 percent for all patents.

In the USA, the period of time from when an application is filed to the moment a patent is issued is 24 months on average. It may take 36 months in the case of biotech and business method patents. Seventy-five percent of applications filed become approved patents. This rate increases if re-filing or sub-division of patent applications into numerous claims is to be taken into account. Trade secrets and patents are often complementary and may be intertwined. Trade secrets may become useful in the early phases of research, before applications for patents are filed, and allow protection of innovations that can be patented only at a subsequent stage. At times, the know-how associated with patents can be protected as trade secrets, but this may be dangerous as it may not comply with patent disclosure requirements. The presence of trade secrets in the first phases of research may hinder the creation of an open research setting in academic and business contexts.

The US patent system is facing major issues related to the greater number of applications and the increased complexity of the inventions. It has to deal with a number of low-quality patents that, despite being limited, is highly costly in terms of resources, uncertainty, bureaucracy, and a reduction in the speed of innovation. Nowadays, careful research is complicated by the high level of complexity of patents and the lack of coordination when applications are examined by the US Patent and Trademark Office (USPTO). The quality of patents assigned is reduced by workforce limitations. Experimental use of patented technologies is being questioned, and this could have a potentially restrictive effect on innovation, which is especially felt in academic and new business environments.

Patenting implies not only advantages but also costs. Monopolies are one of the possible outcomes when a number of patents are grouped by strong companies. In the telecommunications and computing industries, for instance, monopolies may be reinforced and endure because of patents. These may also reduce the creative activity since companies may be less willing to take risks to the point that the total level of innovation may be affected. Moreover, it is necessary to take into account the fact that big companies and independent inventors could have conflicting views and may be concerned about different problems. The two groups will have to find an agreement if a significant change to the patent system is to be made.

There has been an increasing trend over time to reward any type of creativity by protecting IP. As a result, in the broad system of protection only a few domains of free access persist. It is possible to argue that there is no longer an adequate balance between information that is freely available and can be used without cost in the process of invention, and information that is protected and cannot be used without authorization and at no cost. What used to be public research is now becoming proprietary. The pressure of universities toward patenting, and the decline of the public role of corporate research laboratories, are a threat for the public domain and the scientific commons. Too often complex legal procedures determine a clear division between the imaginative process of invention and the benefits of the patent system. A transformation is required in order to limit this influence, so invention can be further encouraged and find more rapidly its way to the marketplace.

INVENTION AND SUSTAINABILITY

Almost three billion people live on this planet with $2 or less per day, and the benefits of human talent do not seem to have affected their lifestyle. Moreover, a number of technologies that have improved our lives are believed to have a negative impact on the environment, causing several problems. Sustainability has become one of the main challenges nowadays. It is fundamental to stimulate the connection between invention and sustainability, but the main issue concerns the way this should be done.

Sustainable development is the means to ensure that present and future generations can enhance their standard of living in an environmentally friendly way. Invention and innovation are key to this practice, and their influence will be even stronger in years to come, especially in developing countries. With regard to the latter, at least three types of invention and innovation can be identified. The first, which can be defined as the "copy-cat," consists in repeating, even without permission, production techniques from developed countries. The second, known as the "piggy-back," consists in the adjustment of existing technologies so they can meet local needs. The third is called the "leap-frog," which consists in the adoption of more sustainable technologies, by excluding those that prove to be inadequate or obsolete. This last type, in particular, is the one that can establish the right connection between invention and sustainability.

In many countries there is a scarcity of resources able to provide a setting in which a creative mind can generate positive outcomes. In particular, very strict and formal education systems tend to repress creativity, especially in developing countries where such systems seem to be the rule. It is also hard for inventors to find the financial resources they need, and also other types of aid, such as mentoring (Greenberg, 2001; Gruber, 1995).

This is especially true for poorer countries due to scarcely developed professional networks. Some countries are still ruled by authoritarian regimes or are dominated by patriarchal social systems. At times, invention is simply not inspired due to the absence of adequate role models. Moreover, in most of the developing countries, invention and innovation are not considered a political priority, and the climate ensuing from this context does not certainly encourage creativity.

The objective of many inventors in poorer countries is not only to produce innovation, but also to create livelihoods in the process of helping communities adopt the new products. Therefore, inventors have a very important social function and are obliged to become social entrepreneurs. Of course, in this setting, they have to face numerous obstacles. Financial institutions are not always willing to provide social entrepreneurs with the resources they need, since they do not have sufficient business experience and do not always value the importance of protecting their inventions; on the contrary, sometimes replication is even fostered, if the inventor believes that by this means his or her product will be used by a greater number of people. Nevertheless, this behavior represents a hindrance impeding social entrepreneurs obtaining the financial resources required by mass production and marketing. In developing countries, inventiveness may be hindered by the new types of IP protection. Patenting is a costly procedure. Furthermore, the sharing of knowledge on sustainable development could be impeded by the existence of patents. Also trade barriers protecting industries in developed countries may prevent livelihoods from being provided in poorer countries. In developed economies the incentives related to inventions specifically addressed to developing economies are very low due to the limited final outcomes of this process in that particular context. For the occurrence of an effective sustainable development, invention should be stimulated through new mechanisms for innovation (Buderi, 2000, 2002; Chandler, 2005). At the same time, manufacturing and marketing systems typically aimed at developing sustainable solutions should be promoted (Geiger, 2004).

Chapter Summary

The high quality of life enjoyed in the developed countries today is mainly due to invention. Of course, not all inventions are advantageous and inequality persists worldwide, with developing countries still not able to fully benefit from invention. However, average living standards in most of the world are much higher than in the past.

At this point, it is interesting to examine the role played by invention and inventiveness in human history. Technological progress has been extremely variable throughout history.

Huge differences among societies can be highlighted with regard to the ability to create and implement inventions. These differences are mainly due to the complex functioning of wider social systems together with their institutions, values, and incentive structures. The inventiveness of a society is fundamentally based on its knowledge base, culture, social priorities, and public policies.

REFERENCES

Ahuja, G., & Lampert, C. M. Entrepreneurship in the large corporation: A longitudinal study of how established firms create breakthrough inventions. *Strategic Management Journal*, Special Issue 22(6–7): 521–543 (2001).

Auerswald, P. E., & Branscomb, L. M. Valleys of death and Darwinian seas: Financing the invention to innovation transition in the United States. *The Journal of Technology Transfer*, 28: 227–239 (2003).

Bozeman, B., & Link, A. N. *Investments in Technology: Corporate Strategies and Public Policy Alternatives*. New York: Praeger (1983).

Buderi, R. *Engines of Tomorrow*. New York: Simon & Schuster (2000).

Buderi, R. The once and future industrial research. In Teich, A. (Ed.) *AAAS Yearbook* (2002)

Chandler, A. *Shaping the Industrial Century*. Cambridge: Harvard University Press (2005).

Darby, M. R., & Zucker, L. G. Grilichesian breakthroughs: Inventions of methods of inventing and firm entry in nanotechnology. *NBER Working Paper No. 9825*. National Bureau of Economic Research, Cambridge, MA (2003).

Dennis, M. Reconstructing sociotechnical order. In Jasanoff, S. (Ed.) *States of Knowledge*. London: Routledge (2004).

Dickson, D. *The New Politics of Science*. Rev. (Ed.) Chicago: University of Chicago Press (1988).

Dickson, D., & Noble, D. The politics of science and technology policy. In Ferguson, T., & Rogers, J. (Eds.) *The Hidden Election*. New York: Pantheon (1981).

Dosi, G. *Technical Change and Industrial Transformation*. London: MacMillan (1984).

Drori, G., Meyer, J., Ramirez, F., & Schofer, E. *Science in the Modern World Polity: Institutionalization and Globalization*. Stanford: Stanford University Press (2003).

Economist Intelligence Unit. *Scattering the Seed of Invention: The Globalization of Research and Development* (2004).

Etzkowitz, H. *MIT and the Rise of Entrepreneurial Science.* London: Routledge (2002).

Etzkowitz, H. Innovation in innovation: The triple helix in university-industry-government relations. *Social Science Information*, 42(3): 293–337 (2003).

Etzkowitz, H., & Leydesdorff, L. The dynamics of innovation: A triple helix of university-industry-government relations. *Research Policy*, 29(2): 109–123 (2000).

Etzkowitz, H., Webster, A., & Healey, P. (Eds.), *Capitalizing Knowledge: New Intersections of Industry and Academia.* Albany: SUNY Press (1998).

Etzkowitz, H., Webster, A., Gebhardt, C., & Terra, B. The future of the university and the university of the future: Evolution of ivory tower to entrepreneurial paradigm. *Research Policy*, (29): 313–330 (2000).

Feldman, M., Link, A., & Siegel, D. *The Economics of Science and Technology.* Boston: Kluwer (2002).

Foray, D. *The Economics of Knowledge.* Cambridge: MIT Press (2004).

Fuller, S. *Thomas Kuhn: A Philosophical History for Our Times.* Chicago: University of Chicago Press (2000a).

Fuller, S. *The Governance of Science.* Buckingham: Open University Press (2000b).

Fuller, S. *Knowledge Management Foundations.* London: Butterworth (2002).

Galison, P., & Hevly, B. (Eds.) *Big Science.* Stanford: Stanford University Press (1992).

Gaudilliere, J-P., & Lowy, I. (Eds.) *The Invisible Industrialist.* London: Macmillan (1998).

Geiger, R. *Knowledge and Money.* Stanford: Stanford University Press (2004).

Gibbons, M. Globalization and the future of higher education. In Breton, G., & Lambert, M. (Eds.) *Universities and Globalization.* Quebec: UNESCO (2003).

Gibbons, M., Limoges, C., Nowotny, H., Schwartzman, S., Scott, P., & Trow, Martin. *The New Production of Knowledge.* London: Sage (1994).

Graham, M. Corporate research and development: The latest transformation. *Technology in Society*, (7): 179–195 (1985).

Greenberg, D. *Science, Money and Politics.* Chicago: University of Chicago Press (2001).

Gruber, C. The Overhead system in government-sponsored academic science: Origins and early development. *Historical Studies in the Physical and Biological Sciences*, 25(2): 241–268 (1995).

Guena, A., Salter, A., & Steinmueller, E. (Eds.) *Science and Innovation.* Cheltenham: Elgar (2003).

Henderson, R, & Cockburn, I. Measuring competence? Exploring firm effects in pharmaceutical research. *Strategic Management Journal*, Winter Special Issue, 15: 63–84 (1994).

Kirchoff, B. Entrepreneurship's contribution to economics. *Entrepreneurship Theory and Practice*, 16(2): 1–14 (1991).

Mokyr, J. *The Lever of Riches: Technological Creativity and Economic Progress.* Oxford: Oxford University Press (1990).

Narin, F., Hamilton, K. S., & Olivastro, D. The increasing linkage between U.S. technology and public science. *Research Policy*, 26(3): 317–330 (1997).

Rubenstein, K. D. Transferring public research: The patent licensing mechanism in agriculture. *The Journal of Technology Transfer*, 28: 111–130 (2003).

Scherer, F. M., & Ross, D. *Industrial Market Structure and Economic Performance.* Boston, MA: Houghton Mifflin (1990).

Schumpeter, J. A. *The Theory of Economic Development.* London: Oxford University Press (1934).

Thursby, J., & Thursby, M. Where is the new science in corporate R&D? *Science*, 314(5805): 1547–1548 (2006).

Zucker, L. G., Darby, M. R., & Armstrong, J. S. Commercializing knowledge: University science, knowledge capture, and firm performance in biotechnology. *Management Science*, 48(1): 138–153 (2002).

CHAPTER 4

OPEN INNOVATION OR COLLECTIVE INVENTION? CONCEPTUALIZING THE DEBATE

Manlio Del Giudice

As Hall, Griliches and Hausman (1986, p. 265) state, "The annual research and development expenditures of a firm are considered to be investments which add to the firm's stock of knowledge." Firms undertake R&D activities, in large part, to create innovations that will ultimately provide new products and therefore profits.

The transition from the industrial society to a knowledge-based economy has been variously described by several authors (Bell, 1973; Gibbons et al., 1994; Hicks and Katz, 1996; Powell and Snellman, 2004; Ziman, 1994). These arguments are very interesting because they show that there has been a definite evolution in the modern research endeavor. There has been a considerable advance in collaboration, at both the domestic and the international level; this has resulted in an increase in the stock of knowledge due to the contribution of different countries and organizations. There has also been a growth in the rate of interdisciplinary research and greater efforts are being made to identify the best applications of basic research in order to find solutions to urgent environmental and health issues. The impact of this change that implies wider collaboration and interdisciplinarity on collective invention is tremendous.

The evolving domain of science can be investigated by utilizing bibliometric evidence, and Hicks and Katz (1996) were among the first scholars to do so. They examined 376,226 publications in the period 1981–1991. Their research revealed that the average number of authors per paper had noticeably increased, from 2.63 to 3.34, while the number of countries and institutions that were represented on each paper had only slightly risen. These discoveries were in tune with earlier research by de Solla Price (1963), which had shown how the relevance of multiple authors had grown over time in areas such as chemistry and physics that the author had described as "Big Science." In more recent times, Wuchty et al. (2007) examined a set of 19.9 million articles and 2.1 million patents related to the period from the late 1950s to the year 2000, and discovered that multiple authors had become increasingly prevalent not only in the field of chemical and physical sciences, but also in the field of social sciences, mathematics, and humanities. As pointed out by Wuchty et al. (2007), in the period 1955–2000, the mean team size at least doubled in the fields of medicine, biology, and physics. The reasons explaining the increase in the number of team members may be related to enhanced knowledge specialization and the increasing costs of doing research. Nevertheless, there has been a growth in the number of authors of papers also in fields where there has been a slower increase in the total number of researchers and the cost factor is not as relevant. Maybe as an inevitable result, Wuchty et al. (2007) discovered, even after double-checking, that papers by teams are quoted more often and the probabilities they may have a strong impact are higher. In later work, Jones et al. (2008) analyzed a set of 4.2 million papers published at US universities in the period 1975–2005, and revealed that authors from several different universities had been increasingly involved by multi-author teams during their work.

If it is true that the total number of inventors grew over time, some important differences deserve careful consideration. Excluding biotechnology, the most marked contrast is to be found in the area of semiconductor-manufacturing processes and pharmaceutical drugs. Semiconductor-manufacturing processes are divided into subsequent phases, which frequently correspond to a specific discipline or the intersection of two disciplines. For instance, several different disciplines are behind most of the current semiconductor manufacturing; optics, chemical engineering, mechanical engineering, and materials science are only a few examples. As pointed out by Orton (2004), every single phase of the manufacturing process, from the design of robotic machines to the chemical baths utilized to eliminate support structures, is a rather distinct body of knowledge. The single phases of the process need to be coordinated, so it is fundamental that the different experts collaborate with one another. Due to the decomposability of the manufacturing process, the teams can remain rather small in size, since there is no need to find the solution to a complex problem at once.

On the contrary, as highlighted by Simon (1962), the field of pharmaceuticals does not frequently offer the possibility to break down problems, because if these were to be decomposed and dealt with separately, the quality of the final outcome would be severely affected.

As pointed out by Arora et al. (2001), drugs are frequently defined by economists as "discrete" technologies because they are not characterized by modularity, while semiconductors and telecommunications equipment are described as "complex" because the high number of parts they are composed of requires integration. Hence, generally, the invention of pharmaceutical drugs cannot be clearly articulated across specific fields of expertise. In most cases, several experts from microbiologists to organic chemists to biochemists, as well as immunologists and pathologists, need to collaborate for the process of invention to be successful. Therefore, the size of the teams depends on the quantity of knowledge that requires integration and the ways in which, thanks to their training and expertise, scientists and engineers manage to deal with technological issues.

Knowledge Sourcing and Costs of Innovation

Technological change can be affected by increasing specialization in two ways. On the one hand, it generates a higher rate of technological progress ensuing from the deepening of knowledge. On the other, it shows that, because of extensive learning and organizational investments, technological progress has become path dependent (David, 1975, 1985; Arthur, 1989; Antonelli, 2007). As pointed out by Patel and Pavitt (1997), as time passes, companies become less willing to change course in their R&D investments. Moreover, as stressed by Cohen and Levinthal (1989), the blinders imposed by their previous work reduce the probabilities that they recognize relevant new knowledge.

As noted by March (1991), and March and Simon (1958), the trend toward local search has long been identified as an issue for all R&D organizations. Several authors have debated around the extent to which learning in research and development is path dependent (David, 1985; Zollo and Winter, 2002), creating the conditions for technological lock-out (Cohen and Levinthal, 1989; Henderson and Clark, 1990; Schilling, 1998). Nelson (1982) believed that the growing depth and extent of potentially significant knowledge has made commercial R&D even more engaging and complex, and it is certainly not a simple research guideline. This challenge is partially related to the problem of finding out which sources of technological opportunity are important and merit to be improved by involving technical staff. Information required to invent new products or processes is provided by sources of technological opportunity (Klevorick et al., 1995; Malerba and Orsenigo, 1997; Cohen et al., 2002).

These have grown in number over time, and are also different from the past from a qualitative point of view:

- Companies obtain their knowledge from more distant geographic areas (see Johnson, 2006; Gittelman, 2007).
- Companies utilize inter-industry knowledge flows more extensively (see Mansfield, 1982; Fung and Chow, 2002).
- Companies tap into a wider range of scientific and technical fields (see Levin et al., 1987; Cohen et al., 2002; Giuri et al., 2007).
- Companies utilize knowledge from universities and government laboratories more extensively (see Powell et al., 1996; Branstetter and Ogura, 2005).

As stated by Antonelli (2001), when latent complementarities are discovered among different resources, collective knowledge comes to light. Nonetheless, since technological opportunities are widely diffused, but the means of attaining returns from innovation are scarce and expensive, the real question relates to how managers decide where to address their search. It is assumed that collective invention is an instrument businesses use to ensure themselves against the risks related to future technologies. As stressed by Powell et al. (2005), collective invention is not only the way to follow new directions, but also allows companies to be "in on the news." Collective invention can also be considered a sort of knowledge insurance for companies that deal with overlapping technical fields. Thanks to knowledge sharing, unexpected technological opportunities may be opened up to businesses. As stated by Cohen and Levinthal (1994), collective invention is a way to access information and develop important skills that assist in adjusting to technical transformation. Issues that had been previously excluded as impracticable are now decomposed thanks to theoretical or technical progress, and wider access to knowledge allows flexibility in factoring complementary advances into research and development (Rosenberg, 1982; Brusoni et al., 2001).

Since the costs of accessing knowledge are high, many companies may be more willing to simply pay the "maintenance costs" of constantly sharing knowledge in collective endeavors.

As pointed out by Cetina (1999), distinct epistemic communities are born because the accumulation of knowledge requires unique theoretical models, and specialized vocabulary and specific software and hardware tools. As stressed by Mokyr (2005), the difficulties in communicating and collaborating widely depend on the epistemic distance between technical communities. Thus, the trend toward localized learning implies that the establishment of a productive dialogue between more distant collaborators will be more complex and time consuming.

Vincenti (1990) argues that, rather than the general learning of new scientific facts or theories, one of the main challenges in the creation of new collaborations is the acquisition of knowledge that is both context and technology specific. According to Nelson and Winter (1982), most of the knowledge of businesses is embedded within routines. Since routines are the idiosyncratic outcome of several historical events, it may be difficult to distribute them in a systematic manner within and across organizations (Von Hippel, 1994; Arora et al., 2001). Kogut and Zander (1993) studied this phenomenon when they investigated 81 cases of technology transfer among companies in Sweden. Respondents were required to give a description of technology being transferred in relation to its codifiability, teachability, and complexity. They were also asked if technology was being transferred to independent companies or fully owned subsidiaries. The scholars discovered that knowledge that was transferred to external companies was generally codifiable and teachable, rather than representing tacit or new ideas. The costs generally related to knowledge transfer, also in the case of joint ventures, in which an intensive collaboration is usually activated, were much higher than in the case of intra-firm transfer.

In order to enhance its internal usefulness, companies tend to use standardized processes and documentation to articulate knowledge; however, according to its tacitness, learning and invention may be required by the process itself (Nonaka, 1994; Nonaka and von Krogh, 2009). The value of knowledge associated with the context has been defined by Von Hippel (1994) as information stickiness. This term is utilized to describe the high costs related to the transfer of knowledge extracted from organizational contexts and routines.

Similarly, as pointed out by Thursby and Thursby (2003), the difficulties connected to knowledge transfer from inventors working in academic environments explain in part why firms cease to work on university-licensed technology. Along the same lines, as reported by Jensen and Thursby (2001), university technologies are more effectively transferred to firms either when they have almost reached the stage of complete development or when they are fully understood by licensing companies. This way, potential incompatibilities between the firm's knowledge base and the university technology are ironed out. Moreover, the commercialization efforts need to be supported by constant academic participation. Hence, knowledge sharing is sped up by formalized knowledge and a common epistemic base. As argued by Rosenkopf et al. (2001), insofar as companies are willing to transfer knowledge from specific sources, it could be generally assumed that collective invention should be encouraged by participation of technical personnel grouped together, either in the form of standard bodies or as communities of practice.

Since technological opportunities may arise from different possible sources and appropriability mechanisms are relatively invariable, simple knowledge of the ways and places to devote research time is a complex matter. Hence, managers have to face the issue of dealing with appropriable short- and medium-term commercial opportunities, while they wait for internal expertise to accumulate through the involvement in activities such as collective invention.

Colocation and Collective Invention

In order to unveil and access complementary knowledge, companies engage in collective invention with remote parties. This is explained by the firms' need to access geographically localized knowledge.

However, for firm formation and innovation a key role is played by colocation (Audretsch and Feldman, 1996; Whittington et al., 2009). For this reason, projects aimed at manufacturing technology products in the short term may be delayed by the distance that separates individuals who have the complementary knowledge required. As highlighted by Breschi and Lissoni (2009), geographic dispersion of knowledge will occur when individuals are not willing to commercialize their technology and publicize their research efforts. As argued by Lakhani (2006), insofar as it is possible to identify a constant diversification in the stock of knowledge, while it is hard to detect complementarity, geographically dispersed accumulation of technological opportunities should increase. Studies on geographically distributed collaboration regarding the past 30 years have revealed that the average distance among collaborators has grown. The reasons explaining this are twofold: on the one hand, access to remote knowledge is increasingly required, and on the other, the constant decrease in the cost of communications technologies has played a significant role. Johnson et al. (2006) analyzed a number of US inventors discovering that between 1975 and 1999 the average distance of co-inventors had increased from 117 miles to almost 200 miles. The authors pointed out that a higher level of clustering could be found in rapidly expanding industries such as computers and biotechnology compared to older ones such as textiles and mechanical devices; however, also the relatively new industries had recently started geographical expansion (Johnson, 2006; Johnson et al., 2006).

Nevertheless, when collaborators are separated by long distances, new difficulties arise. As stressed by Herbsleb et al. (2000), for instance, in commercial software engineering projects, a greater distance involves considerable delays and coordination issues. In their analysis of multidisciplinary National Science Foundation projects, Cummings and Kiesler (2005) discovered that coordination or research achievements are not affected by an increase in the number of disciplinary affiliations; conversely, greater collaboration hurdles are posed by a rise in the number of affiliated institutions. Hence, the most significant challenge for knowledge-based collaborations consists in dealing with coordination difficulties among organizations rather than epistemic distance itself. Since Asian countries, in particular China, Taiwan, Singapore, and South Korea, have published an increasing number of scientific papers, new collaboration and competition patterns able to connect US, EU, and Asian scientists may be needed to face the challenges of distance.

Several scholars have attempted to explain the reasons why distributed collaboration is increasing at a global scale. According to Saxenian and Sabel (2008), it is possible to assume that the creation of specific institutions, such as venture capital, sustains the process of invention by returning immigrants, as strong ties are established with the host country. Saxenian (2006) argues that first-generation immigrants play the role of mediators, since they understand the home nation's culture, and can easily find their way through local institutions, having preserved contacts and relationships. Kerr (2008) analyzed knowledge flows from immigrants back to their home countries, basing his study on variations in US immigration quotas and a scheme that distinguished the names related to diverse ethnicities. Also after verifying how the groups of inventors were composed within specific patent classifications, the scholar discovered that citations were strongly influenced by the community to which the inventors belonged, since the probability that a foreign researcher cites a US inventor of his own ethnicity is 30–50 percent higher. This is particularly evident in the case of Chinese immigrants. As argued by Shrum et al. (2007), in the area of high-energy physics, where it is common that a great number of authors are cited, the standardization of laboratory procedures and well-established conventions regarding experimentation facilitate collaborations among multiple organizations, since they allow extensive teamwork even if the single team members are not well acquainted with one another.

Nevertheless, also in the USA, collaborative and interdisciplinary research does not increase evenly. A study by Jones et al. (2008) on the swift growth of inter-university teams also discovered an increase in stratification. While there had been a rapid expansion of collaboration between single universities, the most influential studies were those in which an elite university had participated. And despite statements by outstanding institutions, such as the National Academy of Sciences (2004), which affirmed the importance of interdisciplinary collaboration in order to "address the great questions of science" and the increasing "societal challenges of our time," the most distinguished universities have been the real protagonists in the construction of interdisciplinary centers. One of the reasons can be found in the capacity of elite universities to attract donations from individuals or organizations that are willing to build those centers. As a result, while research activities expand, social distance persists. As stated by Jones et al. (2008), top universities in the USA are becoming "more intensely interdependent," just when research spreads across organizational and disciplinary boundaries. As a consequence, elite universities have increasingly addressed their research endeavors toward external collaboration, involving in some cases co-participation by business partners. Hence, universities are frequently the basis on which collective invention is built.

A study by Gittelman (2007) unveiled a peculiar trend for papers published by geographically dispersed biotechnology collaborators: these received a high number of citations for their academic work, but they did not garner as many references to their patents. Conversely, more geographically concentrated collaborators were not so frequently cited for their academic work, but received more references to their patents. The author explained these interesting outcomes on the basis of the diversity of the geographic dispersion of knowledge in the cases of public and private science. In Gittelman's view, access to distant knowledge entails costs and benefits, and on this basis it may be assumed that scientifically oriented work is best suited to geographically dispersed teams. Moreover, research at the scientific level more frequently requires codification through formal language, while tacitness is a peculiar feature of work at the engineering level, so colocation is needed in order to convey knowledge from one person to another. These outcomes have significant consequences for collective invention, since its range depends on the tacit or explicit nature of knowledge. In some cases this range may be vast, as for high-energy physics, while in others, such as in a craft-based environment, colocation may be required.

Collective Knowledge, Appropriability, and Technological Regime

One of the topics that has not received due consideration by the literature on collective invention is the shifting focus on innovation and appropriability toward the level of the technological regime rather than the firm. Companies may be more incentivized to make sure that their technological regime is well established so they can take advantage of increasing returns to learning, instead of concentrating their efforts on their technological competitors. A higher specialization and an organizational division of commercial technologies is the result of focusing on appropriability at the level of the technical context. In some regimes there is competition among companies for overlapping intellectual property, but complementary products are manufactured.

In this context, organizations compete for scientific prestige and intellectual property, but in several cases of collective invention, companies do not intend to access the same markets. The stakes of knowledge sharing are much lower since competition among these firms is not driven by the desire to gain market share but is related to the extension of the scope of intellectual property rights. These companies are interested both in the progress of the technical domain and in the additional development of collective knowledge through various outlets. Companies that invest in research and commercialization may be primarily interested in the survival of the technological regime itself, especially when verifying if a technology is commercially viable. In many fields, public research is taking active part in collective invention. The outcomes of government and university research do not generally have an immediate impact on private R&D, with the exclusion of life sciences (Powell et al., 1996; Branstetter and Ogura, 2005; Rhoten and Powell, 2007). A study by Cohen et al. (2002), in which a number of business managers were interviewed, revealed that university and government laboratories were not typically considered as providing direct contributions to the creation of new project ideas. However, the significance of intangible flows of knowledge was stressed by many managers, especially in the form of faculty consulting and contacts at conferences. A research by Branstetter and Ogura (2005) showed that there has been a considerable increase in industry citations to university patents. The result was verified also according to possible variations in the propensity to cite and the available stock of knowledge that could be subject to citation. However, life sciences dominate the increase in the interaction between businesses and universities.

Recently, university–industry licensing has received most of the attention, probably because many universities find it increasingly difficult to access different sources of finance since federal research funds have not grown as much as the costs of research, and businesses only moderately finance basic science (Mowery et al., 2004; Powell et al., 2007).

Certainly, some cases of success in which university licenses have produced considerable revenues have occurred over time. Nevertheless, as pointed out by Zucker and Darby (1996), there is a general imbalance in the distribution of commercial activities by academics. The scholars discovered that more than 20 percent of biomedical publications were produced by a number of star scientists that represented less than 1 percent of the entire population in biomedicine. However, it is believed that, since the extent of these successes is limited, the role played by academic commercial involvement in collective invention is quite confined. Furthermore, successful licenses are frequently based on a simple dyadic interaction between a university and a company, and do not represent a collective license utilized by many.

Nevertheless, as stated by Rosenberg (2000), the needs of businesses are widely met by university research. There are also cases in which universities create a number of complementary inventions according to industry requests. These are incorporated in the academic context and used as research tools or as an engineering system that deserves to be analyzed separately (Rosenberg, 1982; Lenoir and Giannella, 2006). Many aspects can be viewed with regard to the role played by academics in private research and development. Thursby et al. (2009) take account of the cases in which patents were granted by universities to non-academic entities. The scholars discovered that almost a quarter were granted to companies, and faculty consulting is one of the most likely reasons explaining this.

In his research on the tissue-engineering field, Murray (2002) revealed that there are several ways knowledge can exit universities.

Among the most common, it is possible to recall faculty consulting, scientific advisory board memberships, and the exchange of research tools. A research by Fleming et al. (2007) on the inventor networks in Silicon Valley and Boston stressed the key role played by Stanford PhD graduates and a postdoctoral fellowship program at IBM's Almaden Labs, and the relevance of MIT graduates in the Boston community. Whittington (2007) analyzed the inventor networks in life sciences among universities, labs, and firms in the Boston area, and discovered that the key nodes that bound the ecosystem together and made it vital were a few important university laboratories and a number of individual scientists who had moved from universities to companies, or nonprofit organizations to companies, and vice versa. Moreover, universities increasingly take direct part in collective invention. For instance, a repository regarding reusable genetic and proteomic structures has been created thanks to the Biobricks project at MIT. Any company can access the repository and make its contribution in terms of knowledge. At the same time, the licence on the site allows companies to utilize knowledge gained from the repository in order to pursue their business interests. A research by Gittelman (2007) on biotech firm co-authorships revealed that universities or research institutions were research partner of companies in over 90 percent of cases.

A considerable increase in the number of university patents has been pointed out by many. However, discussion arises on whether this is the expression of a growth in valuable applied knowledge or simply a behavior adopted by universities that are attempting to manifest their importance to the private economy (Henderson et al., 1998; Owen-Smith and Powell, 2003; Ziedonis and Mowery, 2004). It is a quite complex task to distinguish the competing influences on universities, but there is no doubt that the number of university patents is increasing.

New Forms of Governance and Collective Invention

Most modern technologies are very complex, and in practice a great number of individuals are required to participate. Thus, collective invention efforts widely rely upon the existence of a social and organizational infrastructure for coordination. Collaboration is also fostered by the shared sentiment that something people will use is being constructed.

According to Wray (2002), scientists and engineers are socialized into rules of collective work because they increasingly rely on common equipment. In general terms, norms and practices of knowledge sharing have been widely refined, since contact across geographic boundaries has been made easier by the development of communications and information technologies (Cummings and Kiesler, 2007; Olson et al., 2008). Here attention is focused on the ways the governance of collective invention is influenced by the use of new collaboration tools, social rules within technical communities, and the organizational characterization of collective invention endeavors.

A technological community is founded on a number of common norms. Hughes (1983) studied the different phases of development of the electric grid in various countries, and argued that every single phase is related to a specific "culture of technology," meaning a series of values and ideas based on which inventors tend toward a common objective. These cultures of technology enable companies to develop a technical endeavor, defined as "technological momentum." According to Mackenzie (1990), technological momentum can be described as an institutionalized type of technological transformation, ensuing from the mobilization of participants around the need to keep a technology alive through alignment of social, political, economic, and technical structures. Institutions are established by individuals not only because they are willing to gain resources and solve technical uncertainties, but also because they have a variety of objectives to align and pursue, and they invest much of their credibility in such a task. The importance of cultures of technology can be also understood in relation to their contribution to explaining the continuity of the technical community that expresses them, regardless of the differences in time required to organize for collective invention versus private research and development.

The historical case of collective invention analyzed by Allen (1983) can be viewed as a place where valuable knowledge is accumulated, or as a supporting community at the crossroads of private interests. When knowledge has accumulated for a while, it can be exploited by internal or external participants through network re-functionality (Carayannis and Campbell 2006). As pointed out by Owen-Smith and Powell (2004) in their study on the development of biotechnology in Boston, in the last two decades of the twentieth century, the community was initially tied to research universities, in particular MIT and Harvard, and also important medical centers such as Massachusetts General Hospital and Dana Farber Cancer Center. These public research organizations were linked to newborn biotech firms through research partnerships and clinical trials. Over the years, venture capitalists appeared, collaborations were established worldwide, and private innovation was attracted by an open community system. Although the imprint of public research persisted, companies tended to be less interested in exploratory research and preferred to forge more product-driven alliances. Leaky ties on which open collaboration had been based were converted into channels of private innovation.

Open innovation and private interest need to be combined in order to generate collective invention. Of course, technical communities can be accessed or abandoned by participants who can exploit their acquaintances for public or private benefit. The most relevant fact, as pointed out by Lakhani and Panetta (2007), in their study on open source, is that

> these systems are not "managed" in the traditional sense of the word, that is, "smart" managers are not recruiting staff, offering incentives for hard work, dividing tasks, integrating activities, and developing career paths. Rather, the locus of control and management lies with the individual participants who decide themselves the terms of interaction with each other.

A study by Hughes (1998) revealed that in the aerospace, computing, and communication industries, technological momentum was gained through the attainment of liquidity and the alignment of political and industrial interests that were behind the systems they produced. For instance, in the case of communications, at a certain point shared objectives were institutionalized through the implementation of standards by the ITU (International Telecommunications Union) that allowed interoperation among regional telephone monopolies. The development of the different technological systems among the geographically dispersed companies required intense coordination, in which systems engineers had an extremely important role. Generally, participation in collective invention is on a voluntary basis, and often inventors can be easily replaced.

Several studies have discussed the reasons why software developers participate to open-source projects. Research by Lakhani and Wolf (2003), based on an Internet survey of 684 software developers across 287 different open-source projects, aimed at explaining community participation. The result was that the most frequent and strong motivation was the simple fact that developers enjoyed their creative work. A secondary motivation consisted in the intellectual challenges related to programming and learning in order to find solutions to users' needs. The creative role played by open-source developers in the creation of effective governance structures is striking. It ensues from the typical interest they have for the work itself and their shared objectives.

As argued by Kling and Iacono (1988), at the group level, computerization is another dynamic that strengthens the impulse to start up and direct collective invention, and it is not the simple outcome of a willingness to gain efficiency. Conversely, the authors claim that the mobilization of participants who support the introduction of information systems is an aspect of computerization of the workplace that has not received adequate consideration. As stated by Fligstein (2001), such a mobilization is achieved by resorting to ideologies that make their way within the organization, but often come from the external setting.

Different kinds of governance are required by different technologies. Collective invention may come to light as a by-product of existing inventive efforts or it may even anticipate the emergence of an organization that exploits the technical knowledge it has accumulated. A wide range of issues can be generally confronted by effective governance mechanisms: coordination problems, compatibility with the rules of specific communities regulating knowledge sharing, and responsiveness to ordinary or complex technical challenges.

As pointed out by Meyer (2003), licenses can be used as a tool to deal with intellectual property barriers to collaboration. In the same line, Gambardella and Hall (2006) believe that the effectiveness of collective invention often requires some type of legal coordination. For example, thanks to the establishment of the General Public License (GPL), future software developers were provided with guidelines according to which they addressed their contributions. Since the most influential developers stressed that their contributions were collective goods, others took the same path and utilized the GPL for the progress of common efforts. In hardware, issues confined to a limited technical space may be solved by resorting to patent pools and cross-licensing; however, when the patents are utilized as a deterrent to new entrants, antitrust concerns may arise.

As argued by O'Mahoney and Ferraro (2007), collective inventors tend to exercise authority through the establishment of formal mechanisms, but use democratic tools to limit their power, thus enabling technical and organizational experimentation. The authors believe that when a common idea of authority is well established, the outcome is generally further extended than the original intent. The systems governing open-source communities have evolved together with common ideas of authority and variable technical goals.

Nevertheless, there can also be coordination without a legal foundation. Over time, researchers have devoted their attention to the fruitful relationships that characterized some communities based on crafts and technology (Scranton, 1997; Sabel and Zeitlin, 1997). For instance, Foray and Perez (2006) highlighted how, in the eighteenth century, the silk industry in Lyon based on open technology had been supported by several political factors. The French region was incredibly vital, and inventors were given grants by the local authorities so the whole community of silk makers could benefit from the sharing of new knowledge. The scholars claimed that even if potential contrasts could have arisen because of collective invention, these conflicts were deadened by general competition and the growth of a sentiment that embraced contribution. As stated by Lamoreaux, Raff, and Temin, in the times preceding the birth of vertically integrated companies, "business people . . . industrial communities interacted socially as well as economically, and the resulting multidimensional relationships facilitated cooperation for purposes besides production."

Along the same lines, research on contemporary high-tech clusters has unveiled different types of private governance that generate collective advantages. The modes common to most prosperous clusters are, in particular, high rates of firm creation, a wide availability of highly qualified technical workers, and inter-firm job mobility (Saxenian, 1994; Bresnahan and Gambardella, 2004). Powell et al. (2009) focused on the emergence of clusters rather than their persistence, in a study on the three US regions that can be considered the cradles of biotech, and in eight areas rich in resources, firms were started up but clusters did not develop. The scholars claimed that participants had to identify new technological paths well before they fully understood their potential. In the case of the biotech industry, anchor tenants, such as venture capitalists in the area of San Francisco or public research organizations in Boston, were of help in such an investigation. They stressed the importance of open science, transparent relationships, and the desire to transfer practices and rearrange them across the public, private, and nonprofit sectors.

Chapter Summary

The key to innovation is an effective management of highly specific knowledge from customers, markets, and other sources. Its significance is shown by high interest in researching on topics such as collective invention or user innovation, just to mention a couple of examples. Nevertheless, numerous difficulties arise when organizations integrate external knowledge in order to encourage innovation. A stimulating approach in current research is the analysis of open innovation as a paradigm shift in the strategic approach incorporating the external world into internal innovation processes.

REFERENCES

Allen, R. C. Collective invention. *Journal of Economic Behavior and Organization*, 4: 1–24 (1983).

Antonelli, C. *The Microeconomics of Technological Systems.* New York: Oxford University Press (2001).

Antonelli, C. Path dependence, localised technological change and the quest for dynamic efficiency. In Antonelli, C., Foray, B., Hall, B., & Edward Steinmuller, W. (Eds.) *New Frontiers in the Economics of Innovation and New Technology: Essays in Honour of Paul A. David.* Northampton, MA: Edward Elgar: 51–69 (2007).

Arora, A., Fosfuri, A., & Gambardella, A. *Markets for Technology.* Cambridge, MA: MIT Press (2001).

Arthur, W. B. Competing technologies, increasing returns, and lock-in by historical events. *Economic Journal*, 99(394): 116–131 (1989).

Audretsch, D. B., & Feldman, M. P. R&D spillovers and the geography of innovation and production. *American Economic Review*, 86: 630–640 (1996).

Bell, D. *The Coming of Post-Industrial Society.* New York: Basic Books (1973).

Branstetter, L., & Ogura, Y. Is academic science driving a surge in industrial innovation? Evidence from patent citations. SSRN eLibrary, http://papers.ssrn.com/sol3/papers.cfm?abstract_id= 788426 (2005).

Breschi, S., & Lissoni, F. Mobility of skilled workers and co-invention networks: An anatomy of localized knowledge flows. *Journal of Economic Geography*, 9: 439–468 (2009).

Bresnahan, T., & Gambardella, A. (Eds.) *Building High-Tech Clusters.* Cambridge, UK: Cambridge University Press (2004).

Brusoni, S., Prencipe, A., & Pavitt, K. Knowledge specialization, organizational coupling, and the boundaries of the firm: Why do firms know more than they make? *Administrative Science Quarterly*, 46: 597–621 (2001).

Carayannis, E. G., & Campbell, D. F. J. *Knowledge Creation, Diffusion, and Use in Innovation Networks and Knowledge Clusters: A Comparative Systems Approach across the United States, Europe, and Asia.* Westport, CT: Praeger Publishers (2006).

Cetina, K. K. *Epistemic Cultures: How the Sciences make Knowledge.* Cambridge, MA: Harvard University Press (1999).

Cohen, W. M., & Levinthal, D. A. Innovation and learning: The two faces of R&D. *Economic Journal*, 99: 569–596 (1989).

Cohen, W. M., & Levinthal, D. A. Fortune favors the prepared firm. *Management Science*, 40: 227–251 (1994).

Cohen, W. M., Nelson, R. R., & Walsh, J. P. Links and impacts: The influence of public research on industrial R&D. *Management Science*, 48: 1–23 (2002).

Cummings, J. N., & Kiesler, S. Collaborative research across disciplinary and organizational boundaries. *Social Studies of Science*, 35: 703–722 (2005).

Cummings, J. N., & Kiesler, S. Coordination costs and project outcomes in multi-university collaborations. *Research Policy*, 36(10): 1620–1634 (2007).

David, P. A. *Technical Choice, Innovative and Economic Growth: Essays on American and British Experience in the Nineteenth Century.* New York: Cambridge University Press (1975).

David, P. A. Clio and the economics of QWERTY. *American Economic Review*, 75: 332–337 (1985).

de Solla Price, D. *Little Science, Big Science.* New York: Columbia University Press (1963).

Fleming, L., King, C., & Juda, A. I. Small worlds and regional innovation. *Organization Science*, 18: 938–954 (2007).

Fligstein, N. Social skill and the theory of fields. *Sociological Theory*, 19: 105–125 (2001).

Foray, D., & Perez, L. H. The economics of open technology: Collective organisation and individual claims in the “fabrique lyonnaise” during the old regime. In Antonelli, C., Foray, D., Hall, B., & Edward Steinmuller, W. (Eds.) *New Frontiers in the Economics of Innovation and New Technology: Essays in Honour of Paul A. David.* Northampton, MA: Edward Elgar: 239–254 (2006).

Fung, M. K., & Chow, W. W. Measuring the intensity of knowledge flow with patent statistics. *Economics Letters*, 74: 353–358 (2002).

Gambardella, A., & Hall, B. H. Proprietary versus public domain licensing of software and research products. *Research Policy*, 35: 875–892 (2006).

Gibbons, M., Limoges, C., Schwartzman, S., Nowotny, H., Trow, M., & Scott, P. *The New Production of Knowledge: The Dynamics of Science and Research in Contemporary Societies.* London: Sage (1994).

Gittelman, M. Does geography matter for science-based firms? Epistemic communities and the geography of research and patenting in biotechnology. *Organization Science*, 18: 724–741 (2007).

Giuri, P., Mariani, M., et al. Inventors and invention processes in Europe: Results from the PatVal-EU survey. *Research Policy*, 36: 1107–1127 (2007).

Hall, B. H., Griliches, Z., & Hausman, J. Patents and R and D: is there a lag? *International Economic Review*, 27(2): 265–283 (1986).

Henderson, R. M., & Clark, K. B. Architectural innovation: The reconfiguration of existing product technologies and the failure of established firms. *Administrative Science Quarterly*, 35: 9–30 (1990).

Henderson, R., Jaffe, A. B., & Trajtenberg, M. Universities as a source of commercial technology: A detailed analysis of university patenting 1965–1988. *Review of Economics and Statistics*, 80(1): 119–127 (1998).

Herbsleb, J. D., Mockus, A., Finholt, T. A.,& Grinter, R. E. Distance, dependencies, and delay in a global collaboration. In *Proceedings of the 2000 ACM Conference on Computer Supported Cooperative Work.* Philadelphia, PA: ACM: 319–328 (2000).

Hicks, D. M., & Katz, J. S. Where is science going? *Science, Technology Human Values*, 21: 379–406 (1996).

Hughes, T. P. *Networks of Power: Electrification in Western Society, 1880–1930.* Baltimore, MD: Johns Hopkins University Press (1983).

Hughes, T. *Rescuing Prometheus.* New York: Vintage Books (1998).

Jensen, R., & Thursby, M. Proofs and prototypes for sale: The licensing of university inventions. *American Economic Review*, 91: 240–259 (2001).

Johnson, D. N. It's a small(er) world: The role of geography in biotechnology innovation. SSRN eLibrary, http://papers.ssrn.com/sol3/papers.cfm?abstract_id¼296813 (2006)

Johnson, D. K. N., Siripong, A., & Brown, A. S. The demise of distance? The declining role of physical proximity for knowledge transmission. *Growth and Change*, 37: 19–33 (2006).

Jones, B. F., Wuchty, S., & Uzzi, B. Multi-university research teams: Shifting impact, geography, and stratification in science. *Science*, 322: 1259–1262 (November 21) (2008).

Kerr, W. R. Ethnic scientific communities and international technology diffusion. *Review of Economics and Statistics*, 90(3): 518–537 (2008).

Klevorick, A. K., Levin, R. C., Nelson, R. R., & Winter, S. G. On the sources and significance of inter industry differences in technological opportunities. *Research Policy*, 24: 185–205 (1995).

Kling, R., & Iacono, S. The mobilization of support for computerization: The role of computerization movements. *Social Problems*, 35: 226–243 (1988).

Kogut, B., & Zander, U. Knowledge of the firm and the evolutionary theory of the multinational corporation. *Journal of International Business Studies*, 24: 625–645 (1993).

Lakhani, K. R. Broadcast search in problem solving: Attracting solutions from the periphery. *Technology Management for the Global Future*, PICMET 6 (2006).

Lakhani, K. R., & Panetta, J. A. The principles of distributed innovation. *Innovations: Technology, Governance, Globalization*, 2: 97–112 (2007).

Lakhani, K., & Wolf, R. G. Why hackers do what they do: Understanding motivation and effort in free/open source software projects. SSRN eLibrary, http://papers.ssrn.com/sol3/papers.cfm?abstract_id¼443040 (2003)

Lamoreaux, N. R., Raff, D. M. G., & Temin, P. Beyond markets and hierarchies: Toward a new synthesis of American business history. *American Historical Review*, 108: 404–433 (April) (2003).

Lenoir, T., & Giannella, E. The emergence and diffusion of DNA microarray technology. *Journal of Biomedical Discovery and Collaboration*, 1: 11 (2006).

Levin, R. C., Klevorick, A. K., Nelson, R. R., & Winter, S. G. Appropriating the returns from industrial research and development. *Brookings Papers on Economic Activity*, 783–831 (1987).

MacKenzie, D. *Inventing Accuracy: A Historical Sociology of Nuclear Missile Guidance.* Cambridge, MA: MIT Press (1990).

Malerba, F., & Orsenigo, L. Technological regimes and sectoral patterns of innovative activities. *Industrial and Corporate Change*, 6: 83–118 (1997).

Mansfield, E. *Technology Transfer, Productivity, and Economic Policy*. W. W. Norton, New York (1982).

March, J. G. Exploration and exploitation in organizational learning. *Organization Science*, 2: 71–87 (1991).

March, J. G., & Simon, H. A. *Organizations*. New York: Wiley (1958).

Meyer, P. B. Episodes of collective invention. *US Bureau of Labor Statistics Working Paper*, http://papers.ssrn.com/sol3/papers.cfm?abstract_id¼466880 (2003)

Mokyr, J. Long-term economic growth and the history of technology. In Aghion, P., & Durlauf, S. N. (Eds.) *Handbook of Economic Growth*. Amsterdam: Elsevier: 1113–1180 (2005).

Mowery, D., Nelson, R., Sampat, B., & Ziedonis, A. *Ivory Tower and Industrial Innovation*. Stanford, CA: Stanford University Press (2004).

Murray, F. Innovation as co-evolution of scientific and technological networks: Exploring tissue engineering. *Research Policy*, 31: 1389–1403 (2002).

National Academy of Sciences. *Facilitating Interdisciplinary Research*. Washington, DC: National Academy Press (2004).

Nelson, R. R. The role of knowledge in R&D efficiency. *Quarterly Journal of Economics*, 97: 453–470 (1982).

Nelson, R. R., & Winter, S. G. *An Evolutionary Theory of Economic Change*. Cambridge, MA: Harvard University Press (1982).

Noble, D. *Forces of Production: A Social History of Automation*. New York: Knopf (1984).

Nonaka, I., & von Krogh, G. Perspective—tacit knowledge and knowledge conversion: Controversy and advancement in organizational knowledge creation theory. *Organization Science*, 20: 635–652 (2009).

O'Mahoney, S., & Ferraro, F. Governance in production communities. *Academy of Management Journal*, 50: 1079–1106 (2007).

Olson, G. M., Zimmerman, A., & Bos, N. (Eds.) *Science on the Internet*. Cambridge MA: MIT Press (2008).

Orton, J. W. *The Story of Semiconductors*. Oxford, UK: Oxford University Press, (2004).

Owen-Smith, J., & Powell, W. W. The expanding role of university patenting in the life sciences: Assessing the importance of experience and connectivity. *Research Policy*, 32: 1695–1711 (2003).

Owen-Smith, J., & Powell, W. W. Knowledge networks as channels and conduits: The effects of spillovers in the Boston biotechnology community. *Organization Science*, 15: 5–21 (2004).

Patel, P., & Pavitt, K. The technological competencies of the world's largest firms: Complex and path-dependent, but not much variety. *Research Policy*, 26: 141–156 (1997).

Powell, W. W., & Snellman, K. The knowledge economy. *Annual Review of Sociology*, 30: 199–220 (2004).

Powell, W. W., Koput, K. W., & Smith-Doerr, L. Interorganizational collaboration and the locus of innovation: Networks of learning in biotechnology. *Administrative Science Quarterly*, 41: 116–145 (1996).

Powell, W. W., Owen-Smith, J., & Colyvas, J. A. Innovation and emulation: Lessons from American universities in selling private rights to public knowledge. *Minerva*, 45: 121–142 (2007).

Powell, W. W., Packalen, K., & Whittington, K. Organizational and institutional genesis: The emergence of high-tech clusters in the life sciences. In Padgett, J., & Powell, W. (Eds.) *The Emergence of Organization and Markets.* Chapter 14 (2009).

Powell, W. W., White, D. R., Koput, K. W., & Owen-Smith, J. Network dynamics and field evolution: The growth of interorganizational collaboration in the life sciences. *American Journal of Sociology*, 110: 1132–1205 (2005).

Rhoten, D., & Powell, W. W. The frontiers of intellectual property: Expanded protection versus new models of open science. *Annual Review of Law and Social Science*, 3: 345–373 (2007).

Rosenberg, N. *Inside the Black Box: Technology and Economics.* Cambridge, UK: Cambridge University Press (1982).

Rosenberg, N. *Schumpeter and the Endogeneity of Technology.* London: Routledge (2000).

Rosenkopf, L., Metiu, A., & George, V. P. From the bottom up? Technical committee activity and alliance formation. *Administrative Science Quarterly*, 46: 748–772 (2001).

Sabel, C. F., & Zeitlin, J. (Eds.) *World of Possibilities: Flexibility and Mass Production in Western Industrialization.* New York: Cambridge University Press (1997).

Saxenian, A. L. *Regional Advantage.* Cambridge, MA: Harvard University Press (1994).

Saxenian, A. L. *The New Argonauts: Regional Advantage in a Global Economy.* Cambridge, MA: Harvard University Press (2006).

Saxenian, A. L., & Sabel, C. Venture capital in the periphery: The new argonauts, global search, and local institution building. *Economic Geography*, 84: 379–394 (2008).

Schilling, M. A. Technological lockout: An integrative model of the economic and strategic factors driving technology success and failure. *Academy of Management Review*, 23: 267–284 (1998).

Scranton, P. *Endless Novelty: Specialty Production and American Industrialization, 1865–1925.* Princeton, NJ: Princeton University Press (1997).

Shrum, W., Genuth, J., & Chompalov, I. *Structures of Scientific Collaboration.* Cambridge, MA: MIT Press (2007).

Simon, H. A. The architecture of complexity. *Proceedings of the American Philosophical Society*, 106: 467–482 (1962).

Thursby, J. G., & Thursby, M. C. Are faculty critical? Their role in university—industry licensing. *NBER Working Paper*, 9991 (2003).

Thursby, J., Fuller, A. W., & Thursby, M. US faculty patenting: Inside and outside the university. *Research Policy*, 38: 14–25 (2009).

Vincenti, W. G. *What Engineers Know and How They Know It: Analytical Studies from Aeronautical History*. Baltimore, MD: Johns Hopkins University Press (1990).

von Hippel, E. "Sticky Information" and the locus of problem solving: Implications for innovation. *Management Science*, 40: 429–439 (1994).

Whittington, K. B. Employment Sectors as Opportunity structures: Effects of Location on Male and Female Scientific Dissemination. PhD Dissertation, Sociology, Stanford University (2007).

Whittington, K. B., Owen-Smith, J., & Powell, W. W. Networks, propinquity, and innovation in knowledge-intensive industries. *Administrative Science Quarterly*, 54: 90–122 (2009).

Wray, K. B. The epistemic significance of collaborative research. *Philosophy of Science*, 69: 150–168 (2002).

Wuchty, S., Jones, B. F., & Uzzi, B. The increasing dominance of teams in production of knowledge. *Science*, 316: 1036–1039 (2007).

Ziedonis, A. A., & Mowery, D. C. The geographic reach of market and non-market channels of technology transfer: Comparing citations and licenses of university patents. In: Cantwell, J. (Ed.) *Globalization and the Location of Firms*. Northampton, MA: Edward Elgar (2004).

Ziman, J. M. *Prometheus Bound: Science in a Dynamic "steady state"*. New York: Cambridge University Press (1994).

Zollo, M., & Winter, S. G. Deliberate learning and the evolution of dynamic capabilities. *Organization Science*, 13: 339–351 (2002).

Zucker, L. G., & Darby, M. R. Star scientists and institutional transformation: Patterns of invention and innovation in the formation of the biotechnology industry. *Proceedings of the National Academy of Sciences*, 93(23): 12709–12716 (1996).

CHAPTER 5

Openness That Matters: Net Generation, Higher Education, and Student Entrepreneurship

Manlio Del Giudice

There is considerable research on "who" is creative, how they think, and the organizational cultures that foster both the generation and the application of creative ideas. Creative leaps, whether small or large, may be thought of as the connection of two or more disparate ideas or concepts within the mind of an individual.

This view of creativity is similar in many respects to the Schumpeterian view of innovation, which consists to a substantial extent of a recombination of conceptual and physical materials that were previously in existence.

Our study begins by pointing to the substantial creativity and innovation benefits available through a network that specifically hones in on knowledge heterogeneity. Knowledge diversity may help students build a sound causal understanding of the relationships between elements in the complex system that they are proposing and so helping them to navigate a project to a successful outcome. Moreover, complex ideas, once generated in the innovating student's mind, must still be understood and supported by others. The nature and probability of success of student entrepreneurial ventures are not simply dictated by personal psychologies, such as tolerance of ambiguity and risk taking, but rather by mobilizing resources to deploy their business plans.

Through exposure to heterogeneous knowledge, for ways of linking students to diverse information to nurture and sustain entrepreneurial activity, the student entrepreneur can draw on expert advocates, that is, contacts who have relevant skills, experience, or know-how to testify to the soundness of particular aspects of a complex new idea, lending their expertise and distinct knowledge to the cause, and so raising its credibility and legitimacy.

In this process, student entrepreneurs use their network relations to access resources necessary for organizational building. The more expertise to which the students have access through their network, the more likely it is that they will be able to amass informed and thus persuasive testimony for their project. Furthermore, novel projects require not only advocacy but also action. Given the inherent uncertainty related to innovation and the difficulty in determining exactly what skills will be needed to move the project forward, having a wide variety of skills on which to draw as needed will also be advantageous. Not much is known about the transition between student network ties and organizational network ties and its effect of new venture performance.

This project investigates the mechanisms that facilitate or impede the transfer of individual network ties to organizational network ties, and how the rate of success of transfer of social capital from individuals to the firm can affect survival and performance of the new venture. It will attempt to bridge the gap between such research literatures and successful student entrepreneurial activity.

It has been widely suggested, and in some respects accepted, that a so-called net generation of students is passing through our universities. Born roughly between 1980 and 1994, these students have been characterized as being technologically savvy, having grown up in an age where computers, mobile phones, and the Internet are part of mainstream culture and society. A number of commentators have even suggested that educators—whom they label "digital immigrants"—need to radically adjust their teaching and learning strategies to accommodate their "digital native" students, predominantly by adopting and capitalizing on the affordances of emerging technologies. Improving students' learning and understanding through the use of educational technologies will depend on a number of factors, including students' motivation and understanding of the potential educational benefits offered by new learning activities. This work aims to explore the notion of the net generation in higher education to gain a better understanding of a range of issues associated with the implementation of emerging technologies in local learning and teaching contexts. We offer our perspective as an orientation to the current state of the art, and we convey our sense of where the stimulating challenges lie.

INNOVATIVE PROCESS AND EXTERNAL KNOWLEDGE

As stated by Cohen and Levinthal (1990), a very significant element in the innovative process is the capacity to gain and utilize external knowledge. In the process of knowledge acquisition, individuals play a crucial role, as different types of external knowledge can be sourced from different individuals. Nevertheless, an in-depth investigation into the subject involved and the types of knowledge pursued has never been carried out. Most research has specifically focused on internal knowledge highlighting the differences between its various types (e.g., tacit and explicit knowledge or codified and noncodified knowledge), while all the types of external knowledge have been generally deemed to have the same significance for an organization. However, the question is whether all the types of external knowledge within the reach of a firm are instrumental for the creation of innovations.

To give a valid response to the argument, it is worth to distinguish between scientific and industrial knowledge. This differentiation may appear quite generic, but it is rather significant as previous investigations pointed out that the nature of scientific and industrial knowledge is diverse: the former is wider and more universal, while the latter is more context specific, idiosyncratic, or problem driven (Allen, 1977; Allen, Tushman and Lee, 1979). A further significant difference is that the aim of scientific knowledge is publicity, while the nature of industrial knowledge is more confidential, since its aim is to take advantage of the profits that may be related to it. The specific features of external scientific knowledge are its wider scope and prompt availability, as well as the higher probability of involving new elements, so, unlike external industrial knowledge, it has a much higher chance of encouraging individual innovativeness.

SOCIAL NETWORKS AND KNOWLEDGE VARIETY

External knowledge is not the sole factor on which comprehension of individual innovativeness is based: the structure of knowledge sharing interactions within a company is equally significant. Social structure gives access to knowledge and information so it is fundamental for individual innovativeness. As pointed out by Burt (1992), brokerage, in particular, intended as the capacity to enter social circles that would differently be detached, has been deemed to play a significant role for the creation of innovations, since it allows to come into contact with non-overlapping information.

Nevertheless, as highlighted by Ahuja (2000), the relationship between brokerage and innovativeness is not very linear for two main reasons: on the one hand, a shared knowledge base is essential for the creation of new ideas founded on external knowledge (Reagans and McEvily, 2003), as knowledge differences are extended through the interaction across social circles and overlapping zones are reduced; on the other hand, as pointed out by Obstfeld (2005), the generation of good ideas is not necessarily the same as the action based on them. It may apparently seem impossible, but if it is true that brokerage gives the variety and complexity of information required to encourage innovation, it is not similarly certain that the social structure beneath it provides the perfect conditions to activate resources, integrate knowledge, and stimulate the convergence of individuals toward new ideas.

This means that there is no evidence that the circumstances offered by brokerage in order to comprehend, share, and profit from the diverse knowledge on which innovations are based are the best possible. In order to solve this issue, it would be preferable that the knowledge of individuals is regarded irrespective of their position in the social structure as a whole (Rodan and Galunic, 2004). In the past, analyses on social networks were generally based on the hypothesis that only by verifying where an individual is situated in the network structure it is possible to assume what is his or her level of knowledge. Consequently, rather than proceeding from actual measurement, knowledge within social networks has been generally deduced from the social structure. For example, similar types of knowledge that overlap can be accessed by individuals embedded in dense redundant structures, while diverse types of knowledge that do not overlap can be accessed by individuals linked to groups that would be otherwise detached.

Nevertheless, it is only possible to hypothesize this type of connection that links a particular social structure to the ensuing knowledge structure. A troublesome issue arises if we deem possible that the knowledge of individuals depends only in part on where they are placed in the social structure, meaning that knowledge in close-knit networks is not completely redundant, and knowledge in sparse networks is partially overlapping.

This assertion is quite sensible because, especially when analyzing innovation, differences in external knowledge originating from outside are a conceivable motive explaining why the knowledge of individuals should not entirely depend on their internal social structure. If compared to internal knowledge, external knowledge is different and heterogeneous, and the diversity in the knowledge derived from outside entails that the knowledge of individuals does not depend on their position in the social structure. If this is true, the capacity of individuals to create innovations is positively influenced by their inclusion in a close redundant network structure.

The various actors will be facilitated in recombining and reassorting their differences in knowledge and views into new ideas, if they manage to continuously interact with each other. External bonds in dense networks would enable new possibilities for the appearance and repetition of different ideas. Moreover, social bonds would have a positive impact on individuals, encouraging them to spend time and energies to interact and transfer knowledge, while, on the contrary, these same possibilities of reiterated interaction and knowledge exchange would not be provided by a disconnected network structure.

Knowledge-Sharing Network and Information-Sharing Relationships

From the discussion above, it can be inferred that the advantages that innovation derives from coordination and collaboration in cohesive networks are fundamentally local, described by the closest circle surrounding the social actors in question.

Nevertheless, the ability to exploit the benefits locally provided by social structures does not cover the entire range of resources that can be accessed in the network, especially if the possibility to acquire information is the main benefit available. Figure 5.1 shows a knowledge-sharing network where the nodes represent individuals and the bonds represent information-sharing relationships.

The graph provides a visual explanation of the distinction between local and global informational advantages. "A" is in a position of high local centrality, if access to new and nonredundant information is considered, because of his links to nonredundant contacts. Though if the entire network is taken into account, his position may be viewed as quite marginal. On the other hand, "B" enjoys not only a high local centrality (consider only the equal amount of nonredundant bonds), but also a high global centrality, while A's global centrality is confined to his surrounding neighborhood. In theory, the difference between local and global informational advantages has consequences on how people look for and discover the information they require.

As previously asserted, the probability of finding new information that could be possibly advantageous arises if there is access to nonredundant contacts. Nevertheless, taking into account the above-mentioned difference, the importance of having nonredundant contacts is certainly great, as the possibility of further extension in

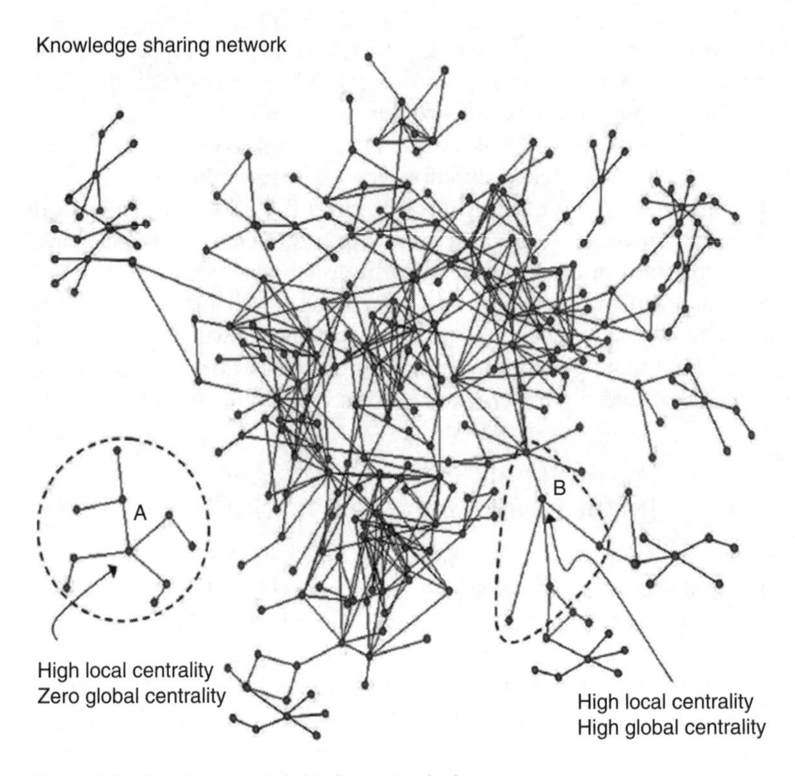

Figure 5.1 Local versus global informational advantages

the global network enhances the probability that new instrumental information may be reached. The possibility of carrying out useful searches proportionally relies on the extent of the reach in the knowledge-sharing network.

THE NET GENERATION IN HIGHER EDUCATION: EARLY THEORETICAL APPROACHES

The concept of net generation has recently gained much interest among scholars who have focused their attention on the evolution of a wide range of people highly skilled in the field of information and communication technologies (ICTs). Net generation members were born in the years from the beginning of the 1980s to the mid-1990s and have been raised in a context in which technology has been extensively used, so they have developed a considerably higher interest for

ICTs than previous generations, and also a greater ability to utilize them (Oblinger and Oblinger, 2005b).

It is general opinion that, with regard to education, net generation students have oriented their preferences and shaped their skills according to their familiarity with information and communication technologies. In fact, they desire to acquire immediate information and process it rapidly, they do not follow a linear course of action when accessing information but they prefer multitasking, they are not passive learners but go ahead and participate actively, and they want to be heard so they constantly interact with others both socially and professionally: thus, communication technologies are fundamental, and experience in such a field is a core part of their education (Barnes, Marateo and Ferris, 2007; Frand, 2000; Gros, 2003; Oblinger, 2003; Oblinger and Oblinger, 2005a; Philip, 2007; Prensky, 2001a, 2001b).

Marc Prensky's (2001a, 2001b) commentaries on "digital natives" and "digital immigrants" have fueled the discussion concerning the educational needs of this generation. Prensky identifies students who attend universities nowadays as digital natives: in his own words, they have "spent their entire lives surrounded by and using computers, videogames, digital music players, video cams, cell phones, and all the other toys and tools of the digital age" (Prensky, 2001a, p. 1). Prensky also states that the amount of digital information these "natives" are exposed to and their way of interacting with it has altered the way they think when compared to previous generations: again using Prensky's own words, "it is now clear that as a result of this ubiquitous environment and the sheer volume of their interaction with it, today's students think and process information fundamentally differently from their predecessors" (p. 1). Prensky also asserts that there is an evident gap between the digital native students and their digital immigrant teachers, due to a significant discrepancy between the culture they both grew up in and the language they use, and this results in a sharp contrast in their skills and preferences. Prensky believes that the greatest issue that education has to face nowadays is the inequality between the ICT skills of today's students and the level of complexity to which such technologies are utilized by teachers. Nonetheless, there has been almost no research comparing the use of ICTs by students and teachers, with the sole exception of the project here presented (Kennedy, Dalgarno, Bennett, Judd, Gray and Chang, 2008).

Prensky's assertions are quite meaningful and require that educational researchers analyze them critically and thoroughly. Nonetheless, as pointed out by Bennett, Maton and Kervin (2008), speculative assumptions have supported much of the discussion regarding the skills and preferences of net generation students relating to ICTs. Although the concept of net generation has been recently examined in many occasions both by the general press and by the academic circles (e.g., Ferrari, 2007; Head, 2007; Leech, 2006; Pesce, 2007), there has not been much investigation regarding the skills and preferences of its members. Nevertheless, this has started to change following the publication of extensive inquiries that show how students use ICTs and other smaller scale surveys regarding what students expect and prefer with regard to the use of digital technology at university.

This research, which is mainly based on data from the United States, points out that youngsters intensively utilize consolidated technologies such as PCs, the Internet, and mobile phones. Other studies investigate the same matter in other countries. For example, data from the Australian Bureau of Statistics (2007) provides general information concerning the use of ICTs, and in particular Internet access, in Australian homes: in 2006–2007, 73 percent of Australian households had a PC and 64 percent had Internet access. Furthermore, with regard to university education, Krause, Hartley, James and McInnes (2005) pointed out that 4.2 hours per week was the average time spent by first-year students on the Internet for study and research purposes, while only 3 percent of students declared that they never used the Internet for their education. In 2007, Oliver and Goerke published the results of their interesting research on first-year students at an Australian university in 2005 and 2007. Most of the students (90 percent) in both years confirmed to have access to the Web outside university, and 93 percent and 87 percent, respectively, asserted they often utilized Internet resources to study. Almost half of the students in the two groups had their own laptops, while only a few had their own tablet computers. Most of the students had mobile phones, and iPods or other similar devices, with a very high increase in percentage from 40 percent in 2005 to 70 percent in 2007. The remaining findings concerned the use of instant messaging (82 percent; 88 percent), blogs (21 percent; 30 percent), and podcasts (7 percent; 22 percent): over the years there had been a constant increase in the use of these communication tools.

Similar research has been performed in the United Kingdom, thanks to the Joint Information Systems Committee (JISC), which funded a number of studies, carried out as part of the "Learner Experiences of E-Learning" project. Qualitative methods were mainly used in order to thoroughly understand the way students considered and utilized ICTs in various educational contexts. Also an online survey was carried out in which the following tools were identified by respondents as those most often utilized for purposes of study and research: precisely, email, the Internet, computers, word processing, and instant messaging (Conole, de Laat, Dillon and Darby, 2006).

In the UK, the JISC studies were more interested in ICTs as learning tools. In the United States, instead, many inquiries provided a much wider view regarding the way American teenagers and college students used information and communication technologies in a more general context. The basis of this research was the PEW Internet and American Life Project (e.g., Lenhart and Madden, 2007), and yearly surveys carried out by the Educause Centre for Applied Research (ECAR). The most recent ECAR report (Salaway, Caruso and Nelson, 2008) was based on a survey of 27,317 students from 98 colleges and universities. Focus group discussions were also performed. The results showed that over 80 percent of respondents had their own laptops and 54 percent owned desktops. Sixty-one percent of respondents declared to have their own mobile devices with Internet access with a considerable increase in percentage compared to the past, even if most respondents stated they did not access the Web via mobile device. Both the 2008 ECAR report and a 2007 PEW report (Lenhart, Madden, MacGill and Smith, 2007) indicate that youngsters intensively utilize social networks, especially Facebook. The ECAR report showed that 85 percent of respondents stated to use social networks, and most of them with the aim of keeping in touch with their circle of friends. The same report also pointed out that other types of technologies were widely used, such as university library websites (93 percent), presentation software (92 percent), spreadsheets (86 percent), text messaging (84 percent), and course management systems (83 percent). Finally, most respondents affirmed that they were expert users of fundamental learning technologies, ranging from "fairly skilled" to "very skilled."

UNORTHODOX LEARNING IN THE DIGITAL AGE

The above-mentioned inquiries prove that youngsters can widely access ICTs and currently utilize them on a daily basis; however, other surveys show a more elaborate view about the students' feelings when using digital technologies for study and research purposes. The most recent ECAR study (Salaway *et al.*, 2008) revealed that just over half of the respondents would prefer a moderate use of information technology when studying, and this is consistent with past ECAR data: in fact, the 2007 report (Salaway, Caruso and Nelson, 2007) showed that students were hesitant and ambiguous in their relationship with technology. For example, if on the one hand most of the students (80 percent) declared to utilize social networks and instant messaging, on the other hand they also admitted they were not willing to avail themselves of these communication technologies for educational purposes, but desired to confine their use within their own private lives. Moreover, respondents stated that in a learning context they appreciated face-to-face interaction with instructors and would never want technology to overshadow this type of personal communication.

The JISC project that aimed at analyzing the way British high school students used ICTs, their attitude toward them, and what they expected from their utilization in an academic context gave similar results (Ipsos MORI, 2007). The project report pointed out that students needed to verify the definite educational and social importance of using digital technology and did not desire to avail themselves of those tools without first recognizing their intrinsic value. The way the students observed in the project understood education, as a conceptualization of the educational process of learning and teaching, affected their consideration of ICTs in an educational setting: "It seems that our audience of young people automatically think of ICT improving their learning through giving them more access to data and research resources, rather than imagining totally new methods of teaching, learning, or interacting with peers and lecturers" (p. 25). Likewise, an ethnographic study by Lohnes and Kinzer (2007) revealed that the students involved in the research had a quite traditional way of considering the educational process, believing that it should mainly consist of a set of activities carried out in a classroom, where learning is basically stimulated by the teacher's skills. Most of the students declared that they normally used digital technologies in their dorms, but were not similarly willing to utilize them in the classroom. One of the students involved in the project liked to use his laptop in class, but the others stigmatized his behavior as antisocial, as they claimed that it created a barrier and disrupted the sense of community within the class. The student using the laptop seemed to have a way of thinking and behaving that is consistent with the general assumptions regarding the net generation. The scholars believed that the results they came to "question the notion that being part of the Net Gen means that college students seek to integrate technology into all aspects of their college experience" (p. 4).

In conclusion, the results from the studies concerning the net generation or digital native students reveal that if, on the one hand, it is true that they have a high level of ICT access and constantly use information and communication technologies, on the other hand, they are less willing to utilize these tools in any setting. For this reason, there could be a discrepancy between what is expected from students by net generation commentators and academic staff about their ICT knowledge and the way they are willing to use it, and their skills and preferences. Thus, as pointed out by Bruns and Humphreys (2007), proper guidance for students is required in order to avoid possible issues when introducing ICTs in a higher learning context, and such guidance should be related not only to the utilization of these technologies, but also to demonstrating how they could promote new ways of learning. These observations are quite significant considering the substantial interest in the potentialities of ICTs in higher learning settings and the idea that students who study nowadays at university are generally expert consumers and creators of Web 2.0 technologies (Bruns, 2007). The next section explains this concept more closely.

PUTTING LEARNING INTO PRACTICE: USING EMERGING TECHNOLOGIES IN HIGHER EDUCATION

As described by Bryant (2007), the term "Web 2.0" refers to Web-based applications including a wide range of Internet technologies, such as social networks, social bookmarking, blogs, wikis, and podcasts. All these various tools are also known as social software, because users do not play a mere passive role as consumers, but actively create content and participate within a community of people, which is the basis of the creation of social networks built using the Web 2.0 tools.

Several commentators have stated that, considering their intrinsic features, Web 2.0 technologies could potentially be powerful educational tools, especially for the net generation (Duffy and Bruns, 2006; Alexander, 2006; Bryant, 2006; Evans and Larri, 2006; Richardson, 2006; Sandars and Schroter, 2007). Nevertheless, as for the opinions regarding net generation students, it is essential that discussions concerning the prospective importance of ICTs in higher learning contexts be based on empirical research that shows how these technologies can be best utilized for educational purposes.

Wikis, podcasts, and blogs are only examples of Web 2.0 technologies extensively used in higher learning settings. Blogging consists in publishing chronologically articles and ideas on a website, where readers can also respond giving their personal contribution to the argument in question (Duffy and Bruns, 2006). Blogging has considerable potentialities as a reflective learning tool encouraging participants to share knowledge. It has been tested in a variety of contexts where this type of tools may be important in an educational perspective, such as in professional development (Instone, 2005), teacher education (Stiler and Philleo, 2003; West, Wright, Gabbitas and Graham, 2006), and business and cultural studies (Williams and Jacobs, 2004; Farmer, Yue and Brooks, 2008). Nevertheless, all these assessments of the importance of blogging in learning settings have suggested that the use of blogging as an educational tool has been variably successful. It has been generally noted that students need to be better directed to the use of blogging in their education, consistently with the different settings in which this tool is embedded (Farmer *et al.*, 2008; Instone, 2005; West *et al.*, 2006).

Similar to blogs, wikis are also starting to be widely utilized in higher education contexts, once again with different levels of success (Bower, Woo, Roberts and Watters, 2006; Bruns and Humphreys, 2005, 2007). Wikis are websites where several users can contribute by editing their content; they are cooperative writing tools, which can potentially build collaborative knowledge. In learning settings, wikis have been used to stimulate interaction among students participating in an online course (Augar, Raitman and Zhou, 2004), build a class-annotated bibliography (Bruns and Humphreys, 2005), develop and publish student essays (Forte and Bruckman, 2006), and support weekly discussion activities and semester-long group projects (Bower *et al.*, 2006).

There are many examples of wikis used in educational settings and their potentialities as learning tools have been widely discussed, but only part of research offers empirical assessments, and the results vary. Bower *et al.* (2006) highlighted a mismatch between the way wikis were viewed by students and staff, the latter being more inclined to avail themselves of wikis to support group work. In the assessment of a wiki as a tool to write and publish essays, Forte and Bruckman (2006) had better outcomes, showing that students were facilitated by the use of the wiki, as they positively responded to the feedback from their peers using the tool. Other research has basically focused on anecdotal evidence, rather than formal assessments, to establish the significance of the use of wikis in higher education contexts (Bruns and Humphreys, 2005, 2007).

Also podcasting has been widely utilized in similar settings. Audio and video educational media have been commonly used for long time, but podcasting and vodcasting are quite recent, and they relate to the diffusion of these types of media via the Internet through a series of syndication feeds to which users are subscribed. On the Web, audio files can be reached by users via streaming or direct download, but without syndication feeds they do not consist in podcasts, which instead are automatically transferred to the users' PCs as soon as they are available, and can be later passed onto a mobile device such as an iPod or a common MP3 player. In higher learning contexts, podcasts have been generally used to deliver lectures and other educational content (Gosper, Green, McNeill, Phillips, Preston and Woo, 2008; Kurtz, Fenwick and Ellsworth, 2007; Lane, 2006; Malan, 2007). Nevertheless, podcasting has also been used to distribute audio recordings created by students as schoolwork, which is definitely more consistent with the idea of Web 2.0 "producers" (Chan, Lee and McLoughlin, 2006; Frydenberg, 2006).

Assessments of podcasting activities performed by students indicate a positive impact on the latter. Research by Chan *et al.* (2006) and Frydenberg (2006) showed that students appreciated very much the experience acquired from the creation of podcasts; moreover, Chan *et al.* (2006) clarified that students who used podcasts considered them useful and educational. Surely students benefit from the possibility of listening to podcasts as they can review lectures when they prefer. In fact, McKenzie's research (2008) showed some interesting results regarding the educational value assigned by students to podcasts, as they believed recorded lectures are a very powerful learning tool, as much as face-to-face lectures. Gosper *et al.* carried out a similar research in an Australian academic context showing that students developed a positive attitude toward the use of lecture podcasts, while university staff was less responsive (Gosper *et al.*, 2008; Phillips, Gosper, McNeill, Woo, Preston and Green, 2007). Nevertheless, other research on the matter in question has produced varying results. For instance, a survey by Kurtz *et al.* (2007) highlighted the students' open hostility toward the use of podcasts, probably due to the replacement of face-to-face lectures as more time was required for group project work in the classroom. Several studies pointed out that students commonly listened to podcasts on their PCs rather than their portable devices, so this questions the idea that one of the most significant benefits of podcasting is the possibility of "mobile ubiquitous learning" (Lee and Chan, 2006, p. 95).

Many scholars have underlined the potentialities of the use of other Web 2.0 technologies, such as social networking, social bookmarking, and digital file sharing for educational purposes (Bryant, 2006; Kamel Boulos and Wheeler, 2007), but little research has been performed in order to assess its significance in higher learning contexts.

EMERGING TECHNOLOGIES AND THE NET GENERATION IN HIGHER EDUCATION: A CONCEPTUAL FRAMEWORK AND PRACTICAL SUGGESTIONS TO STUDENT ENTREPRENEURSHIP

What implications do these perspectives have for the developmental and emergent nature of ICTs in education?

The assessment offered in the literature on the use of emerging technologies in higher education helps researchers define hypotheses that are open to empirical research and identify relevant factors that influence multimodal learning.

A close look at the analytical edifice that is taking shape suggests that there are integrative learning technology dimensions.

Learning by Creating

It has long been recognized in constructivist theories of learning that creating knowledge artifacts is an important element of learning (see Dalgarno, 2001). Developing skills in undertaking independent research is a central component of many higher education courses, and one of the key features of Web 2.0 technologies that aligns well with constructivist views of learning is the possibility for users to create and share content: online publication spaces such as blogs and wikis enable individuals to both read and contribute to the body of information. A number of studies have suggested that students increasingly want to be active participants in the creation of learning content (e.g., Oblinger and Oblinger, 2005; McLoughlin and Lee, 2008). While students have created content as part of their studies for some time (e.g., essays, theses, designs, performances), newer technologies offer the opportunity for student-created content to be easily disseminated among peers, and this sharing of student work is seen as a valuable activity to be used alongside traditional "expert" information sources such as textbooks and lectures.

A primary pedagogical advantage of students creating their own content in learning and teaching contexts is that by creating material, students generate their own internal representations of knowledge consistent with cognitive-constructivist theories of how people learn. And by presenting, articulating, and disseminating their own knowledge and understanding to their peers, students will potentially benefit from a broader conversation about their ideas within the learning community, leading to further knowledge construction and reconstruction consistent with social constructivist theories of learning.

Aside from these advantages, McLoughlin and Lee (2008) argue that by encouraging students to generate content using Web 2.0 technologies, there is great potential to motivate and engage them more fully with their studies, and foster a sense of community. Moreover, it is not unreasonable to expect that by creating content using Web 2.0 technologies, students would develop important generic skills associated with writing, editing, and publishing using different Web-based media

Peer Review for Learning

Social learning theories (e.g., Vygotsky, 1962, 1978) and many contemporary pedagogical models emphasize the important role of interaction, knowledge sharing, and social context in individuals' construction of knowledge and understanding (see Lave and Wenger, 1991; Brown, Collins and Duguid, 1989). Critical components of social learning theories are collaboration and peer-based learning in which students are asked, either explicitly or implicitly, to review and evaluate the work, opinions or ideas, of others. Emerging technologies, particularly social technologies, offer great potential to support this kind of learning activity.

Critical Self-Reflection into Learning Activities

The ability to critically reflect on one's learning is seen by some theorists as an important aspect of student engagement and central to student learning (see, for example, Boud, Keogh and Walker, 1985). There are numerous examples of tools such as blogs that have been employed in higher education with the specific purpose of encouraging students to reflect on their developing knowledge, skills, and experience (e.g., Farmer, Yue and Brooks, 2008; Instone, 2005; Wagner, 2003).

Knowledge Sharing and Collaboration

A key feature of many Web 2.0 technologies is their social nature. Web 2.0 social technologies are able to facilitate collaborative group work (e.g., through a wiki) and file sharing between individuals (e.g., through photo sharing sites such as Flickr, social bookmarking, or podcasting).

Assessment of Student Learning

Designing and conducting assessment that asks students to demonstrate their learning using social Web technologies raises a range of new challenges (Horizon Report, 2008, p. 5). The interactive and creative opportunities facilitated by these technologies may not be used to full effect if academics rely on more traditional, individual forms of online assessment. Crisp (2008) advances design principles for diagnostic, formative, and summative "e-assessment," while Elliott (2007) maps various Web 2.0 authoring forms against different assessment needs. Hughes (2008) sets out criteria for robustness in "e-assessment" that relate to further challenges, such as how to establish academic honesty and integrity and how to manage content for moderation and reporting purposes.

Considering the features of Web 2.0 technologies, several analysts have identified the net generation members as the preeminent users of these tools. Traditional digital communications technologies (mobile phones and email) have recently been supplemented by other Web- and phone-based communications tools, including instant messaging (e.g., *Messenger*) and Web 2.0 technologies such as social networking and blogs. Accordingly, in addition to the more entrenched technologies, this work focused on emerging technology-based tools such as Web-based communications tools including instant messaging and social networking, text-based mobile phone communication, online publishing using blogs and wikis, digital file sharing using the Web and mobile phones, the use of the Web to access published material particularly via syndicated feeds (e.g., RSS) and the use of MP3 players for audio playback and podcasting. The idea that the use of Web 2.0 technologies allows students to produce and not merely consume information is consistent with the outlook of the net generation such as that expressed by Lorenzo *et al.* (2006):

> Constantly connected to information and each other, students don't just consume information. They create—and re-create—it. With a do-it-yourself, open source approach to material, students often take existing material, add their own touches, and republish it. Bypassing traditional authority channels, self-publishing—in print, image, video, or audio—is common. (p. 2)

Online technologies can provide assessment options and opportunities that are simply not possible with traditional teaching and learning methods. Students will therefore need both time and guidance to develop these skills, and educators should allow for this when planning the timing and length of the implementation. This work took initial steps toward addressing such challenges.

STUDENT ENTREPRENEURIAL PROCESS

Earlier studies highlighted the significance of networks and social relationships in the entrepreneurial process, particularly in the start-up phase, as it is possible that they offer the opportunity of accessing many key resources required by the student entrepreneur in the initial stages (e.g., Davidsson and Honig, 2003; Evald *et al.*, 2006; Larsson and Starr, 1993).

Previous research came to the conclusion that possibilities to start up new businesses may be acknowledged by a person who has a higher access level to information because of social networks or differences in search attitude. It is not surprising that the probability of accidentally finding the opportunity of starting up a new business depends on the time spent on search and acquisition of information. Kaish and Gilad (1991) carried out an empirical investigation that supported the relationship between search attitude and opportunity recognition: the authors discovered that the search for information by entrepreneurs mainly followed a nonverbal path and occurred during free time. Regrettably till now, there has not been sufficient investigation on the way student entrepreneurs search for information, especially regarding the behavioral patterns that could be incredibly useful to acknowledge new opportunities.

According to social network theorists, the structure of the social relationships of an individual results in the quantity and quality of information, and the velocity at which he or she can reach the information he or she needs to find out new business opportunities (Aldrich and Zimmer, 1986; Marsden, 1983). Empirical analyses asserting the importance of diverse social networks for the start-up of new enterprises have been recently carried out.

For instance, Renzulli *et al.* (2000) realized that the frequency at which new companies were established by entrepreneurs with links in networks crossing multiple domains of social life was much higher. Similar conclusions were drawn by Burt and Raider (2002), who examined female graduates from a well-known MBA program and discovered that transfer to self-employment was higher among those who were involved in diverse social networks. Nevertheless, while the idea of a higher probability that people with more diverse social networks start up new enterprises is supported by empirical evidence. Stuart and Sorenson (2007, p. 218), regarding the interpretation of a casual connection between diverse social networks and opportunity acknowledgment, stated:

> we should note that most studies of egocentric network structure and entrepreneurial activity examine aggregate data in which the researcher cannot distinguish the network's effect on opportunity identification from its influence on resource mobilization... We consider the evidence to date to fall short of establishing as a stylized fact the idea that diverse networks (those rich in structural holes) enhance opportunity recognition.

As claimed by Burt (1992, p. 47), "contacts strongly connected to each other are likely to have similar information and so provide redundant [informational] benefits," and thus on the contrary it can be inferred that nonredundant knowledge may originate from disconnected contacts.

It is more probable that students involved in a sparse network of detached contacts manage to access a broader range of information regarding the most significant current topics, thanks to its widespread diffusion throughout the network and the possibility to properly verify it through reliable sources.

Although information and knowledge heterogeneity has never undergone direct measurement, it can be well remarked by network structure.

Moreover, the probability that new resources and opportunities are found more rapidly is higher if information and knowledge are diverse: for instance, as pointed out by Granovetter (1974), it is possible to quickly rethink local approaches if promptly grasping the news of evolving strategies on a higher level or again to move faster when new job opportunities appear. A broker can essentially have more complete information if he can acquire more heterogeneous knowledge. In essence, access to more diverse knowledge allows the broker to be informed in a more complete way. As a consequence, it is possible to view a connection linking information and knowledge heterogeneity, which can be reached through a sharing network, and the acknowledgment of effective opportunities. Thus, the first proposition to be reported is the following:

Proposition 1

The acknowledgment of opportunities by student entrepreneurs has a positive connection with the diversity of knowledge existing in the social network in which he participates.

The possibility of accessing diverse knowledge should not only increase the acknowledgment of opportunities, and therefore be related to the capacity of carrying out current activities, but should also enhance the student's potential creativity. In this case, not only the acquisition of normal information regarding news and gossip is involved, but in particular the more profound diversity in the knowledge possessed by the different participants. A creative leap, regardless of its size, can be conceived of as the result of the link between a number of diverse ideas arising in a person's mind (Amabile, 1996; Fiol, 1995; Zaleznick, 1985). If considered this way, creativity is quite similar to the concept of innovation described by Schumpeter, which "consists to a substantial extent of a recombination of conceptual and physical materials that were previously in existence" (Nelson and Winter, 1982; Shane, 2000). For instance, Hargadon and Sutton (1997) took into account a Californian firm that develops new products, IDEO, describing how the company exploits knowledge diversity for the creation of new concepts. It organizes brainstorming activities, bringing together all the teams involved in the different projects with all their heterogeneous knowledge, and gathering a vast number of various artifacts, which are the result of that diversity.

The creative process of generating ideas is stimulated by the recollection, re-assortment, and transfer of the diverse knowledge of the participants that at first sight appears to be disconnected. This integration process is deemed by several scholars as a core cognitive process for the creation of new ideas (Turner and Fauconnier, 1997; Fauconnier and Turner, 1998). The ability to summarize the different information acquired not only relies upon individual skills, but is also affected by the diversity of knowledge people have, which depends on the variety of sources they are in contact with (Carayannis, 2008).

These observations are consistent with Shane's research on entrepreneurs (Shane, 2000), which gave interesting outcomes showing that the probability of success of an enterprise does not merely depend on individual psychologies, such as the attitude toward risk or the acceptance of compromise, but mainly on the knowledge the entrepreneur can preferentially acquire. In essence, the creative process of generating new ideas demands that a person acquires diverse parts of knowledge, probably originating from a social network, in order to recombine them. In fact, much research (Pelz, 1956; Milliken and Martins, 1996; Pelled, Eisenhardt and Xin, 1999) points out that creativity and innovation are also due to knowledge diversity among contacts, which can contribute to the implementation of new ideas, especially when complex tasks are involved. It could allow students to more easily grasp the causal link connecting the different factors in the complex system they are creating and show them the best way to achieve positive results (McGrath *et al.*, 1996). Furthermore, the process of generation of innovative complex ideas does not exhaust in the mind of the innovator, but has to be recognized and accepted by other people in order to be successful. The credibility and legitimacy of a new venture can be supported by the skills and knowledge of third-party experts who can verify and confirm that it is worthwhile to promote the new complex idea generated by the student entrepreneur.

The wider the network of social relations that can support the entrepreneur's idea thanks to the expertise of the contacts involved, the higher the probability that the student entrepreneur will be able to be persuasive and successful. Moreover, innovative projects need not only be supported and promoted, but also demand action. New ideas to be applied carry a great deal of uncertainty along with them, as it is arduous at first to ascertain which skills are mainly required, so the possession of a broad range of abilities to utilize when deemed necessary will also be useful.

In essence, the possibility of carrying out complex tasks will be higher among those student entrepreneurs whose contacts have diverse knowledge and expertise. The opportunity of exploiting heterogeneous skills will be especially advantageous in the case of ineffectiveness in formal procedures to allocate resources, which could be possibly caused by initial indecision regarding those needed to create a new venture.

Thus, the acquisition of heterogeneous expertise and views could be useful to support the start-up phase until the activation of more formal procedures. On this basis, the following proposition can be formulated:

Proposition 2

There is a positive connection between the innovativeness or innovation performance of a student entrepreneur and the diversity of knowledge existing in the social network in which he or she participates.

Furthermore, it is possible to assume that the overall managerial performance and innovativeness are positively influenced by interaction between network structure and knowledge diversity. As stated by Burt (1992, p. 33), "Structural holes are the setting for tertius strategies. Information is the substance." This means that brokering and arbitrage are opportunities offered by tertius strategies, while information is often their bargaining chip. Thus, in the presence of structural holes and the brokering opportunities resulting from them, it can be assumed that the benefits of knowledge diversity are more intense. On the contrary, if there is no heterogeneous knowledge among the contacts through which brokering is performed, the benefits of the structural holes are possibly reduced. Till now, this type of contingent connection has not been verified, as it is probably arduous to empirically analyze the heterogeneity of information and knowledge provided by a social network. This leads to the following proposition:

Proposition 3

There is a positive connection between the overall managerial performance of student entrepreneurs and knowledge diversity and network sparseness acting together.

In the end, to recall Burgelman's ideas on autonomous strategic behavior, it can be stated that innovation is based not only on the production of novel ideas but also on their preservation from those who question them with skepticism. Therefore, for the success of a new entrepreneurial venture, the acquisition and exploitation of heterogeneous knowledge able to bring together new ideas are not sufficient, but also a network structure is required that is useful to support it in the long run in order to make it successful enough to be deemed legitimate. Both are important for the success of an innovation, as the presence of only one of them is not sufficient.

At this point, the social process of innovation can be better understood, as evidence has been provided regarding the role played by the interaction among the different actors involved, external knowledge and the way it circulates, in order to give an account of the ability of student entrepreneurs to give their contribution to the creation of new ventures. The main elements of discussion are the following:

1. The difference between various kinds of external knowledge
2. The difference between social structure and knowledge structure
3. The difference between local and global benefits related to information

If insights concerning the social network theory are considered, it is rather significant to point out that if knowledge of student entrepreneurs is directly regarded without deducing it from the social structure, the way the latter affects innovativeness can be differently understood. Another interesting point of discussion is the difference between local and global benefits related to information; indeed, it is relevant to take into account the broader setting involving local social bonds and not considering only the area that is closest to a student entrepreneur, as much of previous research on social structure has done. The present work has also provided a rather complete literature review, aiming at contributing to the wider field of knowledge management, by showing how discovery, sharing, and readjustment of knowledge within a social network enables to enhance the capacity of a student entrepreneur to create new ideas and give birth to new ventures. One of the most significant outcomes reached in this discussion is that the generation of novel ideas is not the result of external knowledge by itself, but it is the effect of the concurrence of common internal energies aimed at sharing and recombining it.

Chapter Summary

A number of authors have argued that students who are entering the higher education system have grown up in a digital culture that has fundamentally influenced their preferences and skills in a number of key areas related to education. It has also been proposed that today's university staff are ill equipped to educate this new generation of learners—the net generation (the so-called group of school children dubbed "digital natives")—whose sophisticated use of emerging technologies is incompatible with current teaching practice. The purpose of this work is to provide a balanced exploration of research and practical implications on the impact of digital technologies on higher education to make sense of the dynamics we observe.

The study points to the substantial creativity and innovation benefits available through a network that specifically hones in on knowledge heterogeneity. Our concern for both knowledge and networks led us to introduce a unique way of using network methods to study the knowledge diversity of interpersonal networks. Finally, our study should be of considerable practical concern for student entrepreneurship.

References

Ahuja, G. Collaboration networks, structural holes and innovation: A longitudinal study. *Administrative Science Quarterly*, 45: 425–455 (2000).

Alexander, B. Web 2.0: A new wave of innovation for teaching and learning? *Educause Review*, *2006*(March/April): 33–44 (2006).

Aldrich, H. E, & Zimmer C. Entrepreneurship through social networks. In Sexton, D. L., & Smilor, R. W. (Eds.) *The Art and Science of Entrepreneurship*. Cambridge, MA: Ballinger: 3–23 (1986).

Allen, T. J. *Managing the Flow of Technology*. Cambridge, MA: MIT Press (1977).

Allen, T. J., Tushman, M. L., & Lee, M. S. Technology transfer as as function of position in the spectrum from research through development to technical services. *Academy of Management Journal*, 22(4): 684–708 (1979).

Amabile, T. A. *Creativity in Context* (2nd edn). Boulder, CO: Westview Press (1996).

Augar, N., Raitman, R., & Zhou, W. Teaching and learning online with wikis. In Atkinson, R., McBeath, C., Jonas-Dwyer, D., & Phillips, R. (Eds.) *Beyond the Comfort Zone: Proceedings of the 21st ASCILITE Conference* (pp. 95–104). Perth, Australia. Available from: http://www.ascilite.org.au/conferences/perth04/procs/pdf/augar.pdf (2004).

Barnes, K., Marateo, R. C., & Ferris, S. P. Teaching and learning with the net generation. *Innovate, 3*(4): (2007).

Bennett, S., Maton, K., & Kervin, L. The 'digital natives' debate: A critical review of the evidence. *British Journal of Educational Technology, 39*(5), 775–786 (2008).

Boud, D., Keogh, R., & Walker, D. (Eds.) *Reflection: Turning Experience into Learning*. London: Kogan Page Ltd. (1985).

Bower, M., Woo, K., Roberts, M., & Watters, P. *Wiki Pedagogy—A Tale of Two Wikis*. Paper presented at the International Conference on Information Technology Based Higher Education and Training (ITHET) (2006).

Brown, J. S., Collins, A., & Duguid, P. Situated cognition and the culture of learning. *Educational Researcher*, 18(1): 32–42 (1989).

Bruns, A. Produsage: Towards a broader framework for user-led content creation. In Shneiderman, B. (Ed.) *Proceedings of the 6th ACM SIGCHI Conference on Creativity & Cognition C&C '07* (pp. 99–105). New York: ACM Press (2007).

Bruns, A., & Humphreys, S. Wikis in teaching and assessment: The M/Cyclopedia project. In Riehle, D. (Ed.) *Proceedings of the 2005 International Symposium on Wikis*. New York: ACM Press: 25–31 (2005).

Bruns, A., & Humphreys, S. *Building collaborative capacities in learners: The M/cyclopedia project revisited*. Paper presented at the Wiki Symposium 2007 (2007).

Bryant, T. Social software in academia. *Educause Quarterly* (2), 61–64 (2006).

Bryant, L. Emerging trends in social software for education. In *Emerging Technologies for Learning: Volume 2*. Coventry, UK: Becta: 9–18. Available from http://emergingtechnologies.becta.org.uk/index.php?section=etr&catcode= ETRE_0001&rid= 14167 (2007).

Burt, R. S. *Structural Holes: The Social Structure of Competition*. Cambridge, MA: Harvard University Press (1992).

Burt, R. S., & Raider, H. J. Creating Careers: Women's paths to entrepreneurship. Unpublished manuscript, University of Chicago, Chicago (2002).

Carayannis, E. G. Conceptual framework for an analysis of diversity and heterogeneity in the knowledge economy and society. In Carayannis, E. G., Kaloudis, A., & Mariussen, Å (Eds.) *Diversity in the Knowledge Economy and Society: Heterogeneity, Innovation and Entrepreneurship*. Cheltenham, UK: Edward Elgar Publishing: Ch. 5, pp. 95–116. (2008).

Chan, A., Lee, M. J. W., & McLoughlin, C. (2006). Everyone's learning with podcasting: A Charles Sturt University experience. In *Proceedings of the 23rd Annual Ascilite Conference: Who's Learning? Whose Technology*? The University of Sydney. Available from http://www.ascilite.org.au/conferences/sydney06/proceeding/pdf_papers/p171.pdf

Cohen, W. M., & Levinthal, D. A. Absorptive capacity: A new perspective on learning and innovation. *Administrative Science Quarterly*, 35: 128–152 (1990).

Conole, G., de Laat, M., Dillon, T., & Darby, J. (2006). *JISC LXP: Student Experiences of Technologies: Final Report:* Joint Information Systems Committee. Available from http://www.jisc.ac.uk/

Crisp, G. *Raising the Profile of Diagnostic, Formative and Summative e-assessments: Providing e-assessment Design Principles and Disciplinary Examples for Higher Education Academic Staff.* ALTC Fellowship Report. http://www.altc.edu.au/carrick/go/home/pid/342 (accessed on February 1, 2009) (2008).

Dalgarno, B. Interpretations of constructivism and consequences for computer assisted learning. *British Journal of Educational Technology*, 32(2): 183–194 (2001).

Davidsson, P., & Honig, B. The role of social and human capital among nascent entrepreneurs. *Journal of Business Venturing*, 18(3): 301–331 (2003).

Duffy, P., & Bruns, A. The use of blogs, wikis and RSS in education: A conversation of possibilities. In *Proceedings of the Online Learning and Teaching Conference 2006*. Brisbane, Australia: 31–38. (2006).

Elliott, B. *E-assessment: What is Web 2.0?* Glasgow: Scottish Qualifications Authority. http://www.sqa. org.uk/sqa/22941.html (accessed on February 1, 2009) (2007).

Evald, M. R., Klyver, K., & Svendsen, S. G. The changing importance of the strength of ties through the entrepreneurial process. *Journal of Enterprising Culture*, 14(1): 1–26 (2006).

Evans, V., & Larri, L. J. *Networks, Connections and Community: Learning with Social Software*. Canberra: Australian Flexible Learning Framework (2006).

Farmer, B., Yue, A., & Brooks, C. Using blogging for higher order learning in large cohort university teaching: A case study. *Australasian Journal of Educational Technology*, 24(2): 123–136 (2008).

Fauconnier, G., & Turner, M. Conceptual integration networks. *Cognitive Science*, 22(2): 133–187 (1998).

Ferrari, J. Go digital or lose out, teachers told. *The Australian*, p. 9, November (2007).

Fiol, M. C. Thought worlds colliding: The role of contradiction in corporate innovation processes. *Entrepreneurship Theory and Practice*, 19(3): 71–91 (1995).

Forte, A., & Bruckman, A. From Wikipedia to the classroom: Exploring online publication and learning. In Barab, S., Hay, K., & Hickey,

D. (Eds.) *Proceedings of the 7th International Conference on Learning Sciences*. Bloomington: Indiana: Indiana University: 182–188 (2006).

Frand, J. L. The Information-Age mindset: Changes in students and implications for higher education. *Educause Review*: 15–24 (2000).

Frydenberg, M. Principles and pedagogy: The two P's of podcasting in the information technology classroom. In *The Proceedings of ISECON 2006* (Vol. 23). Dallas (2006).

Gosper, M., Green, D., McNeill, M., Phillips, R., Preston, G., & Woo, K. *The Impact of Web-Based Lecture Technologies on Current and Future Practices in Learning and Teaching*: *Australian Learning and Teaching Council*. Available from http://www.altc.edu.au/carrick/go/home/pid/347 (2008).

Granovetter, M. *Getting a Job* (2nd edition). University of Chicago Press: Chicago, IL (1974).

Gros, B. The impact of digital games in education. *First Monday*, 8(7): (2003).

Hargadon, A., & Sutton, R. I. Technology brokering and innovation in a product development firm. *Administrative Science Quarterly*, 42: 716–749 (1997).

Head, B. The D Generation: Leading the emerging generation of digital natives calls for a light touch. *AFR Boss*. August, 26–30 (2007).

Horizon Report. Stanford, CA: New Media Consortium/EDUCAUSE Learning Initiative. http://www.nmc.org/pdf/2008-Horizon-Report.pdf (accessed on February 1, 2009) (2008).

Hughes, J. Open accreditation—a model. *Pontydysgu—bridge to Learning Blog*. http://www.pontydysgu.org/2008/10/open-accreditation-a-model/ (accessed on February 1, 2009) (2008).

Instone, L. Conversations beyond the classroom: Blogging in a professional development course. In *ASCILITE 2005: Balance, Fidelity, Mobility: Maintaining the Momentum?*: 305–308. Available from http://www.ascilite.org.au/conferences/brisbane05/blogs/proceedings/34_Instone.pdf (2005).

IPSOS MORI. *Student Expectations Study: Key findings from Online Research and Discussion Evenings held in June 2007 for the Joint Information Systems Committee*: JISC. Available from http://www.jisc.ac.uk/(2007).

Kaish, S., & Gilad, B. Characteristics of opportunities search of entrepreneurs versus executives. *Journal of Business Venturing*, 6(1): 45–61 (1991).

Kamel Boulos, M. N., & Wheeler, S. The emerging Web 2.0 social software: An enabling suite of sociable technologies in health and health care education 1. *Health Information and Libraries Journal*, 24(1): 2–23 (2007).

Kennedy, G., Dalgarno, B., Bennett, S., Judd, T., Gray, K., & Chang, R.. Immigrants and Natives: Investigating differences between staff and students' use of technology. In *Hello! Where are you in the Landscape of Educational Technology? Proceedings Ascilite Melbourne 2008*. Available from http://www.ascilite.org.au/conferences/melbourne08/procs/kennedy.pdf (2008).

Krause, K., Hartley, R., James, R., & McInnes, C. *The First Year Experience in Australian Universities: Findings from a Decade of National Studies.* Canberra, Australia: Department of Education, Science and Training (2005).

Kurtz, B. L., Fenwick, J. B. J., & Ellsworth, C. C.. Using podcasts and tablet PCs in computer science. *Paper Presented at the ACM South East Regional Conference* (ACMSE 2007), Winston Salem, NC, March 23–24 (2007).

Lane, C. *UW Podcasting: Evaluation of Year One.* Washington: Catalyst Office of Learning Technologies (2006).

Larsson, A., & Starr, J. A. A Network model of organization formation. *Entrepreneurship Theory & Practice*, 17(2): 5–15 (1993).

Lave, J., & Wenger, E.. *Situated Learning: Legitimate Peripheral Participation.* Cambridge; New York: Cambridge University Press (1991).

Lee, M. J. W., & Chan, A. Exploring the potential of podcasting to deliver mobile ubiquitous learning in higher education. *Journal of Computing in Higher Education*, 118(1): 94–115 (2006).

Leech, R. Teaching the digital natives. *Teacher: The National Education Magazine*, 2006, March 6–9 (2006).

Lenhart, A., & Madden, M. *Teens, Privacy & Online Social Networks: How Teens Manage their Online Identities and Personal Information in the Age of MySpace.* Washington: Pew Internet & American Life Project (2007).

Lenhart, A., Madden, M., MacGill, A. R., & Smith, A. *Teens and Social Media: The Use of Social Media Gains a Greater Foothold in Teen Life as they Embrace the Conversational Nature of Interactive Online Media.* Washington: Pew Internet & American Life Project (2007).

Lohnes, S., & Kinzer, C.. Questioning assumptions about students' expectations for technology in college classrooms. *Innovate*, 3(5), http://www.innovateonline.info/index.php?view= article&id= 431.14 (2007).

Lorenzo, G., Oblinger, D. G., & Dziuban, C. *How Choice, Co-Creation, and Culture Are Changing What It Means to Be Net Savvy* (No. ELI Paper 4): Educause (2006).

Malan, D. J. *Podcasting Computer Science E-1.* Paper presented at the SIGCSE. March 7–10, Covington, Kentucky, USA (2007).

Marsden, P. V. Restricted access in networks and models of power. *American Journal of Sociology*, 88(4): 686–717 (1983).

McGrath, R. G., Tsai, M.-H., Venkataraman, S., & MacMillan, I. C. Innovation, competitive advantage and rent: A model and test. *Management Science*, 42(3): 389–403 (1996).

McKenzie, W. Where are audio recordings of lectures in the new educational technology landscape? In *Hello! Where are you in the Landscape of Educational Technology? Proceedings Ascilite Melbourne* (2008).

McLoughlin, C., & Lee, M. J. W. Mapping the digital terrain: New media and social software as catalysts for pedagogical change. In *Hello! Where are you in the Landscape of Educational Technology? Proceedings Ascilite Melbourne 2008*: 641–652. Available from http://www.ascilite.org.au/conferences/melbourne08/procs/mcloughlin.html (2008).

Milliken, F. J., & Martins, L. L. Searching for common threads: Understanding the multiple effects of diversity in organizational groups. *Academy of Management Review*, 21(2): 402–433 (1996).

Nelson, R. R., & Winter, S. G. *An Evolutionary Theory of Economic Change*. Cambridge, MA: Harvard University Press (1982).

Oblinger, D. G. Boomers gen-xers millennials: Understanding the new students. *Educause, 38* (July/August) (2003).

Oblinger, D. G., & Oblinger, J. L. Is it age or IT: First steps toward understanding the Net Generation. In Oblinger, D. G., & Oblinger, J. L. (Eds.) *Educating the Net Generation* (pp. 2.1–2.20) EDUCAUSE. Available from www.educause.edu/educatingthenetgen/ (2005a)

Oblinger, D. G., & Oblinger, J. L. (Eds.) *Educating the Net Generation*: EDUCAUSE. Available from www.educause.edu/educatingthenetgen/ (2005b)

Obstfeld, D. Social networks, the tertius iungens orientation, and involvement in innovation. *Administrative Science Quarterly*, 50: 100–130 (2005).

Oliver, B., & Goerke, V. Australian undergraduates' use and ownership of emerging technologies: Implications and opportunities for creating engaging learning experiences for the Net Generation. *Australasian Journal of Educational Technology*, 23(2): 171–186. (2007)

Pelled, L. H., Eisenhardt, K. M., & Xin, K. R. Exploring the black box: An analysis of work group diversity, conflict, and performance. *Administrative Science Quarterly*, 44(1): 1–28 (1999).

Pesce, M. Brace for a steep re-learning curve. *The Age*, December 2 (2007).

Pelz, D. C. Some social factors related to performance in a research organization. *Administrative Science Quarterly*, 1(3): 310–326 (1956).

Philip, D. The knowledge building paradigm: A model of learning for Net Generation students. *Innovate*, 3(5): (2007).

Phillips, R., Gosper, M., McNeill, M., Woo, K., Preston, G., & Green, D. Staff and student perspectives on web based lecture technologies: Insights into the great divide. In ICT: *Providing Choices for Learners and Learning. Proceedings Ascilite Singapore 2007*: 854–864. Available from http://www.ascilite.org.au/conferences/singapore07/procs/phillips.pdf (2007)

Prensky, M. Digital natives, digital immigrants. *On the Horizon*, 9(5) (2001a).

Prensky, M. Digital natives, digital immigrants, Part II: Do they really think differently? *On the Horizon*, 9(6): (2001b).

Reagans, R., & McEvily, B. Network structure and knowledge transfer: The effects of cohesion and range. *Administrative Science Quarterly*, 48(2): 240–267 (2003).

Renzulli, L., Aldrich, H., & Moody, J. Family matters: gender, networks, and entrepreneurial outcomes. *Social Forces*, 79: 523 (2000).

Rodan, S., & Galunic, C. More than network structure: How knowledge heterogeneity influences managerial performance and innovativeness. *Strategic Management Journal*, 25: 541–562 (2004).

Richardson, W. *Blogs, Wikis, Podcasts, and Other Powerful Web Tools for Classrooms*. Thousand Oaks, CA: Corwin Press (2006).

Salaway, G., Caruso, J. B., & Nelson, M. R. *The ECAR Study of Undergraduate Students and Information Technology, 2007.* Boulder, CO: Educause Center for Applied Research. Available from http://www.educause. edu/ecar (2007).

Salaway, G., Caruso, J. B., & Nelson, M. R. *The ECAR Study of Undergraduate Students and Information Technology 2008.* Boulder, CO: Educause Center for Applied Research. Available from http://www.educause.edu/ecar (2008).

Sandars, J., & Schroter, S. Web 2.0 technologies for undergraduate and postgraduate medical education: An online survey. *Postgraduate Medical Journal*, 83: 759–762 (2007).

Shane, S. Prior knowledge and the discovery of entrepreneurial opportunities. *Organization Science*, 11(4): 448–469 (2000).

Stiler, G. M., & Philleo, T. Blogging and blogspots: An alternative format for encouraging reflective practice among preservice teachers. *Education*, 123(4): .789–797 (2003).

Stuart, T. E., & Sorenson, O. Strategic networks and entrepreneurial ventures. *Strategic Entrepreneurship Journal*, 1(3–4): 211–227 (2007).

Turner, M., & Fauconnier, G. A mechanism of creativity. *Poetics Today*, 20(4): 397–418 (1997).

Vygotsky, L. S. *Thought and Language* (E. Hanfmann & G. Vakar, Trans.). Cambridge, MA: MIT Press (1962).

Vygotsky, L. S. *Mind in Society*. Cambridge, MA: Harvard University Press (1978).

Wagner, C. Put another (B)log on the wire: Publishing learning logs as weblogs. *Journal of Information Systems Education*, 14(2): 131–132 (2003).

West, R. E., Wright, G., Gabbitas, B., & Graham, C. R. Reflections from the introduction of Blogs and RSS Feeds into a preservice instructional technology course. *TechTrends*, 50(4): 54–60 (2006).

Williams, J. B., & Jacobs, J. Exploring the use of blogs as learning spaces in the higher education sector. *Australasian Journal of Educational Technology*, 20(2): 232–247 (2004).

Zaleznick, A. (Ed.) *Organizational Reality and Psychological Necessity in Creativity and Innovation*. Ballinger: Cambridge, MA (1985).

Chapter 6

On the External Dimension of Business Knowledge Flows: "Markets for Knowledge Resources"

Maria Rosaria Della Peruta

To understand the phenomenon of *markets for (knowledge) resources*, which have determined a progressive and marked widening of the area of sourcing knowledge resources for business, it is necessary to fix, as a starting point, analyses of business behavior (Pierce and Aguinis, 2013). These analyses show, in an unequivocal manner, how the search for original techniques for the management and development of knowledge constitutes a fundamental element in the competition between organizations (Leiponen and Helfat, 2010; Ireland et al. 2009; Foss and Klein, 2012). If we concentrate on the logic of strategic and organizational behavior able to test the solutions and settings through which businesses reveal their own potential in order to attain competitive advantage, the alternative of access to knowledge outside the confines of the business cannot be overlooked (Pérez-Luño et al. 2011; Mol and Birkinshaw, 2009). It returns powerfully to represent a valid move for businesses that are operating in contexts of high complexity, contexts where the processes of adaptation to technological, market, and competitive forces are extremely difficult (figure 6.1).

Markets of knowledge resources foresee various transaction typologies through which the use, diffusion, and creation of technological knowledge occur. The transactions can foresee entirely technological packets (patents, other intellectual property, and know-how), or transactions that regard non-patentable knowledge or knowledge without patents (for example, software or numerous non-patented projects and innovations). In this text the term "market" shall be used in a wide sense. In rigorous terms, market transactions are easily made, are anonymous and entail a typical exchange of goods for money. Many, if not most, transactions of technology fall short of these criteria. Often these transactions have very detailed contracts and may be included in some type of technological alliance. The way technology is commercialized is connected to its particular nature as an economic resource and potential object of exchange. Technology appears in various forms and no generic definition is adequate. For example, it may take the form of "intellectual property" (like patents) or as intangible goods (for example, a software program or a project) or again it may take the form of technical assistance.

Figure 6.1 Markets of knowledge resources

Although, from 1970 to the present, there has been surprising growth in the sum total of the research and development activity produced *extra moena*,[1] the phenomenon of external sourcing is relatively recent. Trying to use an interpretative lens of a historical evolutionary nature, one sees the trend toward a progressive return to what already existed at the beginning of the twentieth century, when the positioning of R&D laboratories inside the business was considered an isolated strategic choice. From the increase in awareness that monitoring of the trajectory of technological development and the speed of change required knowledge and expertise, which that a business rarely managed to possess or to elaborate in conditions of efficiency, the proliferation of contracts stipulated with universities and independent scientists, recorded from the first decades of past century,[2] demonstrated the irreversibility of a process turned toward not only operative, but also strategic interdependence.[3]

It was only in the period of the First World War (for the USA) and even more so in the Second World War (for Great Britain and for many leading European countries[4]) that businesses in the more industrialized countries started to privilege internalization of all R&D activity as a crucial choice for industrial competitiveness in economies increasingly based on the generation and use of new knowledge.[5]

Very soon, the complex conditions of the competitive contests that characterized the last decades of the twentieth century made activation of a latent need for independence and autonomy at the origin of entrepreneurial actions almost unrealistic and proposed new dynamics for the accumulation and sharing of knowledge with a reduction of costs and risks connected to innovation. In fact, the current phenomenon of *external sourcing*, unlike the traditional version, widespread until the middle of the plast century, has been accompanied by a continuous growth of technical capacity and internal research, a strategic process that also considers the need for businesses to keep their own *core* technological expertise,[6] to adequately evaluate their own capacities, and to select and then use the technical elements and research brought in, with a capacity for absorption which conditions the rate of growth.

In fact, we live in a new, different "era" in terms of research expertise and technological abilities and in collaborative models for businesses. This new type of business, with innovative behavior and operational modes, looks for a more mature and balanced approach to satisfy technical and research needs;[7] if, on one hand, contravening with respect to the traditional mode of development, they "borrow" resources from other businesses and in parallel "lend" or concede resources and knowledge to other businesses, even the competition, on the other hand, they keep investing in research and development to increase the capacity of absorption and integration of their businesses, that is, a greater maintenance of the relevant information so as to make use of it.

Respect for such a condition allows businesses to develop a greater *absorptive capacity*[8] through which knowledge taken from outside is better assimilated to be introduced into the business on a hierarchic basis, instead of being exclusively contractual. The strategic objectives that are expected to be obtained are not, in any case, different to those achievable through direct investments and internal development, but they substantially change the methods used, the delivery times, and the barriers to imitation; the presence and the creation of alternatives to be adopted widen the possibilities for growth of innovative capacity and, by extension, the ability to generate knowledge in the future.

Transaction Costs, Capabilities, and Vertical Scope

Our research initiatives regarding *markets for knowledge resources* start, as shown in the interpretative proposal advanced in the preceding paragraph, from a concept of business, which connects development of its own resources to a process of growth in which it depends on other businesses, with which it interacts and exchanges, to gain advantage from this collective effort for the production of new knowledge.[9]

This way of reinterpreting a business leads to highlighting how the limits to development, to paraphrase Penrose, are no longer represented by growing costs or the lack of opportunity, but on the capacity or lack of capacity for managing the fragmentation and distribution of the knowledge necessary for operating successfully in environments continually modified by innovation, competition, and institutional change.[10]

In the activity of businesses, historically, the vision of a growth process made by recurring to the market, rather than by the internal development of resources, is not very lucid (and scarcely planned). On the other hand, the incomplete and imperfect character of the markets, in which the environment is not sufficiently wealthy in resources (or not of all the necessary resources) and in transparency regarding relevant information (given the different players involved), does not favor a structural dependence to profit from external resources.

Despite this, if we concentrate on cases where in which these resources are available and accessible with coherent governance methods for the relationships (buying and selling) between businesses, then it is possible to make a choice, and, when managed in a programmed fashion, the way a business concretizes this may represent a great opportunity.

Success or failure may depend on such choices. In the technological sector, the decision to go it alone can slow down the development of the final product to such a point that the entire initiative fails; on the other hand, drawing technology[11] of fundamental importance from the market may lead to *royalty* costs and technological restrictions able to hamper efforts toward future development and equally condemn the project to failure.

The chance of giving yourself the advantage of multiple strategic alternatives, undeniably, calls for careful reflection on markets and hierarchies as emerging reference points in the relationships between organizations for the production of innovation. In fact, the difficulties inherent in the separation of activities regarding the development of innovation[12] have favored the integral management of "knowledge" and slowed down the emergence of markets for knowledge resources.

Teece[13] confronts this problem claiming that the interdependence between the tasks in the process of innovation and the natural uncertainty associated with the development and commercialization of innovation create three principal sources of transaction costs. The first source relates to the difficulty in providing detailed specification of the needs of the tasks at the beginning of the innovative process. These specifications can be exactly determined only with the start of the process, which requires largely incomplete contracts, and leaves ample space for the possibility of opportunistic behavior by others. If, later, a business develops difficult relationships with the technology supplier, the disagreement can generate *sunk costs*, which most probably lead to an increase in the *switching costs* and the problem of *"lock in."* Finally, freeing precontractual information in favor of those offering technology (*bidders*) means sharing this precious information and increasing the risk that competitors can come to know about the R&D plans.

These considerations allow us to understand why, traditionally, the planning, development, and commercialization of new products and processes has have remained integrated within businesses.[14] Vertical integration, which evokes a form of hierarchy and self-sufficiency of the business, helps to resolve the problems linked to opportunistic behavior, which arise when the contracts are still incomplete. In the case of innovation this means that the business must specify and organize the actions of different players involved in the innovative process while the process itself is beginning, especially in the case of more complex technologies, like systems, that clearly require interdependence between the actors.[15]

Because of the high transaction costs connected to the complexity and interdependence of the innovative phases, which can hinder businesses in the separation of one or more activities, like production and commercialization, many businesses have had to renounce the possibility of exploiting extended economies of scale in the sector and in the market of production technology and R&D, exploiting only limited economies at the level of the business.

In fact, for deciding on the opportunity (or not) of integration (in a historical context), businesses should confront their *abilities* with those of other businesses, represented by the conditions of price and quality under which the latter are disposed to distribute their resources.

These simple observations constitute a fundamental point that most economic analyses, and in particular the theory of transaction costs (TCE), avoid considering for analytic convenience; in fact, TCE concentrates on the conditions of exchange and ignores the conditions of production (Demsetz, 1988; Coase, 1988, 1992; Langlois and Foss, 1999; Riordan and Williamson, 1985). This implies that it is necessary to look at the distribution of production capacity—the source of every potential profit deriving from exchanges in the market (Teece,

1980; Jacobides and Hitt, 2004)—to understand when businesses are integrated and when they are not.

Market transactions occur when both parties (in the exchange) find the specialization sufficiently advantageous to overcome all the exchange costs; this implies particular properties in the distribution of capacity of the sector. In particular, if the productive capacities, at a sector level, are symmetrically distributed in the segment "upstream" and "downstream" and if there are no capacity limits or bottlenecks, then no benefit is to be had from the exchange (*intermediate trade*) (Jacobides, 2004). If, however, the capacities are there but are weakly correlated between the various stages—the efficiency of the downstream segment translates, in an approximate way, to the upstream segment—then specializations will be followed if the transaction costs allow.

It seems reasonable to suppose that such a correlation (weak, or even negative) will be the rule when the productive capacities in the different parts of the value chain isare built on different knowledge bases. In other words, the more dissimilar the segments, the less the force of one translates into the force of the other, the less the possibility that a single business can claim efficiency upstream and downstream (Jacobides and Winter, 2005).

When the level of analysis moves from the individual business to the population of businesses involved in the market of knowledge resources, it becomes clear that vertical specialization must be partly a function of heterogeneity in the productive capacity along the "*cumulative chain of innovation*." The transaction costs play a role similar to that of a tax on sales or a tariff imposed on international commerce. If the asymmetries in the production capacity of the innovative process are high, such "taxes" may not be sufficient to limit vertical specialization. A business thinking of "using" the market might be disposed to pay (even reluctantly) these high "taxes," whether they are a *frictional cost* or costs for the predicted losses of *hold-up*,[16] renegotiation, etc., if the profits deriving from the exchange, due to the heterogeneous productive capacity, were high enough to compensate for these losses.[17]

Such theoretical deductions evoke the classical image of the efficient business, committed to maintaining its own boundaries and disposed to sacrificing itself to the market, to an efficient market where the latent profits induce change in the *vertical scope* of the existing businesses, influencing the entry conditions for new businesses and the emergence of new typologies of business.

In fact, "it is the chance of making profits which cannot be captured within the existing *arrangemental structure* which leads to the formation of new (or to the mutation of old) methods through which the parties govern the exchange (*institutional arrangements*)" (Davis and North, 1971, p. 39). This is the conceptual base—and the most widely shared theoretical conviction—which evokes the classical image of the economic institutions, "firms as nexus of contracts" in the words of Jensen and Meckling (1976), which evolve toward efficient economic solutions through negotiations between the interested parties (*contracting*): as Demsetz (1967) suggested, as long as the cost–benefit analysis indicates potential profits, one shall see a change in the system of *property rights*, which brings about those potential profits, through variations (*adjustments*) in the market price and in the production capacity to which the existing institutions of the governance of transactions (*institutional arrangements*) are harmonized.

IPRs and Income from Innovation

The traditional analyses of *property rights* (Demsetz, 1967; Davis and North, 1971; North and Thomas, 1973) paint an optimistic perspective that presupposes the emergence, over time, of an economically efficient system of *property rights*, aimed at "internalising the externalities."

One may reflect on the following example. Suppose that a producer, henceforth the buyer B, requires a specialist input. In addition, we assume that this input represents the unique *specialty* of the scientists and of engineers, who in turn have a *research unit*, offering the supply S. One way for B to obtain the input is to *employ* "group S"—vertical integration. Alternatively, the members of S could constitute an independent business (business S) to offer the input to B through contractual forms (*arm's-length contracts*).[18] The " preference of many engineers and scientists to work in smaller organisations"[19] could bring about greater efficiency,[20] in that, in a separate business the esprit de corps of group S would not be weakened by the presence of an *overarching corporate structure*, with its inevitable bureaucratic costs: on the other hand, integration could bring about significant benefits, including greater information control, which, for business B, is better put into effect through relationships with personnel than through having the status of simple contractor[21] as occurs, in fact, in relationships that are generated between a mother organization and a spinoff business. More generally, we may say that, compared to an integrated entity, two independent businesses would be less efficient in coordinating the exploitation of information *spillover*. In other words, integration avoids the dissipation of the profits from information *spillover*.

Naturally, with integration business B sacrifices any strong incentives toward contractual forms.[22] A contract could better motivate group S to use its talent and its specific knowledge in an efficient way for adapting the input to the needs of the business B. In the absence of an efficient contract,[23] however, group S would need to be better protected so as to protect the investments made in the effort of adaptation.[24]

In particular, regarding the technology markets,[25] an interesting analysis is made by Arora and Merges, who have modeled the *make or buy* decision of a business as a tradeoff between integration, a choice that avoids the dissipation of income (for the *spillover* of the benefits of R&D), and a *freestanding input supply firm*, a choice that takes advantage of the specialization (because of the incentives linked to contractual forms through intellectual property rights[26]).

The model assumes the existence of *special capabilities* as the endowment of group S, which in turn could become either a division of B or a separate business S. From the point of view of business B,[27] the decision on externalizing knowledge—that is, making a contract with business S—depends, in part, on the capacities developed *intra moena*.[28] Business S, on its part, as a separate business, has a potential performance advantage on group S, division of B, but if the *special capabilities* of group S are strongly correlated ("sticky") to the complementary resources of business B, the force of *property rights* could encourage group S to invest in separating ("unsticking") its own know-how from the context of B. In other words, the capacity of setting up group S as business S may depend on the ability of S to anticipate expropriation by B, using determined mechanisms.[29]

The problem of "theft" (*the "hold-up" problem*), posed by the holders of external patents, which is seen when one party is able to expropriate income from another, is similar to that highlighted in the literature on transaction costs (see, e.g., Williamson, 1985; Klein et al., 1978).[30] The public nature of the intellectual property and the uncertainty of the costs regarding delimitation of the boundaries of ownership lead to an important point in this typical *hold-up problem*. If believed valid, a patent confers on its owner, for a limited period, the right of excluding others from the possibility of using the patented invention. However, it does not give the right to the patent holder to use the patented invention if such use violates the rights of others (*it is an exclusionary, not an affirmative right*).

If a business creates an invention autonomously and uses it to improve the quality of its products or its production methods, that business does not necessarily have the right to "apply" or use its invention if that could lead to a violation of the rights of owners of other patents. According to the frequency of simultaneous use and of the reproduction of inventions, a business finds itself in the position of deciding whether to "create and buy" or "create or buy" in the knowledge market. Thus, attention moves to the price that a business anticipates paying in the case of a need to buy the legal rights to be able to use technologies patented by others and the ways of improving its own contractual position ex post. In theory, a business could simply invent some modifications to the technology in possession of others, reducing, in this way, the risk of committing a potential theft. On this issue, hypotheses on the timing of investments, on their practicality, and on the costs of ex ante contracting are of crucial importance.

Consider, for example, the problem from the point of view of a producer of semiconductors: the producer could easily invent modifications on an existing patent, for example, during the initial phases of planning new products or when new fabrication plant becomes available. In this case, the royalties that the holder of the patent that has been modified could obtain from the latter would be limited, ex ante, by the ability of the producer to think up modifications on the invention (Levin et al., 1987; Teece, 1986). However, the producer would find themselves in a decidedly disadvantageous contractual position when it became aware of the patent only after having incorporated the technology. At this point, the invention represents a highly specific resource (in the classical sense of transaction costs) even in the case in which the identity of the holder of the resource were unknown before making the decision to invest.

Compared to the theoretical tradition (Libecap, 1989), more recent research stresses how the external award of property rights influences the practicability of coming up with ex ante solutions (see, e.g., Heller, 1998, 1999; Heller and Eisenberg, 1998), in that conceding too many individual exclusion rights (compared to an excessively limited scale) can, in fact, impede the use of economic resources. As a consequence, the resource can be underused. A subtle but important concept follows from this "anti-commons" or "diffuse entitlements theory," which is that the contractual challenge of a business is influenced by the level of dispersion among rights holders, not only by the patents "jungle" or the number of holders (per se) (as indicated by Shapiro, 2001).[31]

If, on the one hand, intellectual property rights are believed to provide an ex ante incentive to innovate, even if this also comes at a cost of limiting the spread of the technology, on the other hand, greater protection of IPRs could be socially undesirable ex post.

In reality, for the purpose of our work, what is interesting to stress is that the IPRs englobe the codification of information, which contributes to "separating" that information from the context of origin and making its transfer simpler, thereby determining the locus, and not simply the level, of investments and incentives to innovate; the allocation of *property rights* plays a determining role in influencing the decisions on the boundaries of the business and in making the contractual context more dynamic.

Internal Development Versus External Sourcing: The Perspective of a Resource-Based View

The possibility of taking advantage of the existence of markets in knowledge resources provides absolutely unique ways for businesses' strategic moves.

When a resource is available through the relative markets, businesses can choose to constitute reserves through internal accumulation, external availability, or, as generally occurs, through a combination of both.

Making a choice between internal and external availability of resources (*internal and external resource sourcing*) is very different from the *make or buy* decision for production. The decision of external sourcing of resources also relates to the prices of the components supplied, the delivery programs, and the transaction costs.

Where the entrepreneurial component, in terms of motivation and investment behavior, takes on substantial relevance in terms of competitive advantage, the cost parameter loses significance. In particular, the observation of the moves of businesses that have used development routes centered on external sourcing of knowledge resources shows that actual performances are connected not only to cost differentials but also to different ways of generating profits, which concern questions like the governance regime of intellectual property rights, the cover of the patents of the resources, and the terms for which these may be transferred.

For a business that undertakes an initiative of *external knowledge sourcing*, the questions focus on various aspects: discovering where to go to find those resources (a sort of *competitive intelligence*), making an agreement that guarantees the effective transfer of the resources (as well as the contractual details of the agreements, for example, licences), learning and assimilating externally sourced resources, and managing the relevant inter-business relationships.

These are the problems that arise from the choice of external sourcing, which are not normally taken into consideration in the conventional analysis of accumulation of resources for maintaining competitive advantage, but are widely analyzed in the literature on organizational agreements to obtain profits deriving from innovation (Teece, 1986, 1992, 1998). From now on, such arguments shall be encompassed by the term *strategic acquisition of knowledge resources* to establish that this type of process has a status of equal importance to that of internal accumulation of resources, which truly contribute to the process of accumulation of resources when businesses are able to look beyond themselves to build up their own reserves of "knowledge."[32]

The parallel process for businesses that are making dynamic modifications to their reserves of resources, for example, when they change their strategic priorities, *is a strategic dismissal of resources.* Such a process includes cases where the businesses dismiss operative plant, or some division, for what is generally described as a *market for corporate control*, but in reality is another form of market of resources. As a consequence, the strategic dismissal of resources allows businesses to thin down their own reserves in conformity with the current strategic plans (Mahoney and Pandian, 1992): this is the twin process that uses the dynamic of resources to reinforce the industrial dynamics and *sectorial patterns* of innovation (Malerba and Orsenigo, 1997).

The acquisition of strategic resources, together with strategic dismissal, contributes to the dynamic adjustment of the reserves of the resources of a business on the basis of its own strategic objectives of the moment, as hypothesized by Grant (1991) in his concept regarding the dynamic process of adjustment of resources. According to Grant, this dynamic adjustment of resources is a way of making up for an absence of current resources. Grant does not enter into details about the way in which this gap is effectively bridged. Warren (1999a, 1999b, 2002) further develops this concept of dynamic adjustment of resources and discusses the relevant processes, focalizing attention, in particular, on questions of complementary nature of resources and feedback.

The concept of strategic acquisition of knowledge resources, which highlights different phases of sourcing and obtaining resources, as well as governance of the inter-business relationships relevant to them, in some way leads to a model that expresses the idea that the creation and maintenance of competitive advantage emerges from the capacity of a business to own and manage, in detail, adequate solutions with respect to the need to create value in turbulent contexts.

In fact, an examination of the processes of strategic acquisition of resources verifies that businesses are involved in a dynamic process of adjustment of reserves of resources in such a way that these reserves can be rebuilt and redimensioned, through the acquisition or the dismissal of resources, according to the required strategies. Businesses use to the acquisition of external resources to support their own efforts toward internal development, overcoming the restrictions of a program with long and uncertain times and utilizing integration of resources that give them the capacity and experience able to favor organizational solutions and to create conditions of "context" for a strategic combination of new resources and knowledge. Incumbent businesses are called upon to decide (1) which of the innovative capacities and the technological competencies they do not have must be bought; (2) which research fields not yet explored must be investigated; and (3) which, among the selected research lines, must and can be followed within their organizational boundaries.

Discussion of the main problems linked to the theme of maintaining competitive advantage through innovation, in the presence of a market of resources where businesses exchange and share resources in an evolutionary and co-evolutionary dynamic, represents an effort in interpreting emerging entrepreneurial attitudes adopted in the development of innovation, in light of the "resource based view" (RBV), in which the definition of the boundaries of R&D and the structure governing the activity takes on the value of a fundamental choice for the development and use of the business knowledge base.

As is noted, RBV claims that to maintain a higher-than-average source of income, these resources must satisfy three criteria: that is, to be of value, rare, and not perfectly mobile (Barney, 1991; Peteraf, 1993; Markides and Williamson, 1996). In other words, a competitive advantage must be sustained by resources for which there are not, and there cannot be, well-functioning markets. Businesses construct sustainable competitive advantages through access to resources that the competition cannot access. Barney (1986) observes that the possession of these resources must be based on imperfections of market factors, that is to say, a market where factors used to create these resources are commercialized. By definition, these imperfections are triggered by the fact that businesses have differing expectations with respect to the future value of these resources (Barney, 1991). Cool and Dierickx[33] (1989) claim that not all the resources needed to maintain a competitive advantage can be bought and sold. On the contrary, these resources must be accumulated within the business through determined mechanisms defined over a certain period of time. In the same way, much of the reflection on technological strategy has addressed the problem implicitly or explicitly accepting that technological resources cannot be bought and sold directly, nor can their services be "hired." What happens when some of these previously non-tradable resources become so?

By contrast, here, the acquisition of external resources is presented as a rational and calculated strategic action that integrates the internal processes of development. In a similar context, the normative implications of traditional RBV are insufficient (Teece, Pisano, Shuen, 1997). Although, in fact, they represent the natural theoretical catchment area within which to study the methods of formation of strategies, given the substantial inapplicability of the prescriptions of a structuralist type in highly dynamic contexts, they require substantial integration.

While the orthodox *economic discourse* of RBV, which employs the method of balances, allows identification of properties that resource assets must possess to generate value, the attempt that is posed here is to direct research toward an understanding of the methods of business growth on the basis of their own "excess" of resources, and new behavioral and entrepreneurial models at the base of dynamic competitiveness.

The critical process of the creation of new resources, which directs economic knowledge and adjustment, that is, evolutionary success, may have origins, for example, in the development of new technological standards that guide the production of a new sector, as occurs, for example, in the pharmaceutical industry with the rise of biotechnologies; businesses need to manage the diversity of the knowledge structures, in an emerging technological regime, so as to favor the adoption and use of a set of technologies that influence the development of the products, also with regard to the technical knowledge that shall become dominant in the sector.

"Resource Dynamics" and the Evolutionary Approach

The starting point for the application of RBV in the extended form is consideration of the way in which resources can be contained within the businesses and how the latter can produce income of a Ricardian or Schumpeterian type with such resource packets. By Ricardian incomes we mean profits obtained by exploiting the rarity and the superiority of the resources of a business, as well as the peculiarity of the procedures created so as to be able to use them. Schumpeterian incomes are profits of an entrepreneurial type, which the business draws from resource packets in synergy among themselves and coming from a variety of sources.

This leads to questioning what determines the growth rate of businesses considered as "resource packets," the limits of this growth, the circumstances that bring businesses to dismiss resources, and lastly the way these aspects are translated into entrepreneurial and managerial practices. While businesses transform their recently discovered activities into something "routine," managerial attention is free to address further discoveries, thereby allowing businesses to grow, diversify, and build on the basis of their own "excess" resources, that is, on the co-specialization of resources that act in synergy among themselves. Businesses expect complementary resources from those businesses with which they have a direct relationship, through the dynamics of diffusion, reproduction, speculation, and transfer of resources. These aspects represent the exchange dynamic of the resource economy, guided by considerations about the disequilibrium (rather than considerations about the equilibrium that control the neoclassical analyses of the economy of goods and services), which takes businesses toward new behavioral and entrepreneurial models.

Considering it from the point of view of resources, the concept of co-specialization of resources, be it within a business or between businesses, may be interpreted as an expression of co-evolutionary dynamics, which, in turn, reinforces knowledge in the economic field. If the resources can be described in terms of their evolutionary and co-evolutionary dynamics, what, then, is the significance of this point of view for economic performance?

Variety inside the economic system increases the capacity of businesses to choose and acquire resources for the constant reconfiguration of the existing assets and gives value to efforts toward the creation of those naturally complex *capabilities*, aimed at integrating and coordinating forms of specialist knowledge.

In biological evolution, one frequently observes the phenomenon of species co-adaptive to changes in their environment, so that they co-specialize with respect to each other. This event is defined as co-evolution. Numerous examples include micro-organisms that evolve in the intestines of certain species of mammals or ants that co-evolve with certain types of acacia to provide reciprocal advantages. Now we observe that businesses also function according to co-evolutionary principles.

In its extended form RBV gives rise to important intuitions in the area of the evolutionary and co-evolutionary dynamics of the economy, which are based on variations, selection, and conservation of the resources within businesses and between them. The elements that constitute the evolutionary approach to the economy are, today, sufficiently well defined. Nelson and Winter (1982) first gave a clear summary of evolutionary theory, posing it as an alternative to the traditional neoclassical theory of the economy, which is more static and aimed at optimisation. The authors proceed by making reference to businesses (understood as "phenotypes") and their organizational routines (understood as "genes" or "genotypes") considered as continuity factors for economic life, unlike the casual fluctuations and the reactions tending toward optimization with regard to prices, both foreseen in the neoclassical point of view.

The RBV developed in this study can take on the definition given by Nelson and Winter, further elaborating it and specifying that it is not only about "routine," but rather about resources that act as units of variation, selection, and conservation. The RBV, looking at the "economy as a whole," gives a description that unifies economic evolutionary processes through the dynamics of variation, selection, and conservation of resources.

The Darwinian synthesis, in the way it is used today, makes a fundamental distinction between *replicators*—where a variation is seen—and *interactors*—where there is a selection.

The dominant concept at the basis of the viewpoint of replicators–interactors is that the evolutionary process proceeds through the variation of the replicators, which offer particular advantages (or disadvantages) to the players who have the roles of interactors. In the world of biology the replicators represent the genetic material, while the interactors are the organisms. In the cultural and behavioral world, instead, the replicators are memories while the interactors represent the people whose brains contain these memories. In the business world, the replicators are the business resources, the routines and the relationships created between them, while the interactors are the businesses themselves. These businesses are then selected through market competition, to then copy the replicators that have given them the advantage. A particular aspect of the evolutionary model is the need of a perspective of reproduction rather than the usual company perspective, since the dynamic of the system is observed through the work of these two levels and the processes that interconnect them.

Hence, four arguments present themselves for immediate analysis: (1) How are the variations carried out in the course of reproduction of resources, routines, and relationships? (2) How do the competitive prospects of a business depend on these variations? (3) How are efficient variations transmitted? (4) How is knowledge connected to evolutionary development?

Above all, we consider the reproduction of resources, routines, and inter-business relationships. The resources are reproduced when businesses that develop them, or more frequently a competitor business, try to recreate resources using both tacit and explicit elements. Competition through imitation is a very powerful motor of the economic dynamic. The routines are reproduced when a business newly creates routines it has had access to and applies them to its own activity.

This also provides another privileged point of view for examining the large-scale diffusion of economic resources and routines. Businesses constantly seek to inform themselves of the routines of other businesses so as to reproduce them within their own activity. Relationships between businesses are reproduced when a business moves toward a new area of activity and recreates the connections it had previously interwoven with other businesses. The simplest example is the case of producers in the Japanese automobile sector, which move overseas. They gain advantage from the fact that they can draw on a system for the reproduction of relationships. The variations in these processes of reproduction can be propagated through the economy and lead to differential selection pressures that are tried by businesses that incorporate the variations into resources, routines, and inter-business relationships.

At this point, we turn our attention—as in the formulation of the evolutionary theory—to considering the resources, routines, and relationships as replicators, that is, as elements that propagate independently of the desires of businesses that generate them in the first place. The key question is interpreting how variation within the economic system—which is the key of adaptive responses to changes in external conditions—may be generated by variation in the replicators that are at the base, in this case considered as resources, routines, and inter-business relationships. Variation is considered Darwinian, in the sense of its pure variety and in the sense that even though variations are purposely introduced by businesses, their consequences within a wider economy cannot be foreseen (Mathews, 2003).

The evolutionary perspective that has contaminated the orthodox version of theoretical studies of RBV makes a clear review of the strategic behavior that was latent for a long time; in light of this fertile contamination, a reference to strategy refers not only to the cognitive ability of top management and to its capacity of taking the "right" decisions (Burgelman, 1994), but also to the capacity to work creatively with primary materials, the "right knowledge" offered both by the environment and by business (Quinn, 1978; Mintzberg, 1987), and finally to the capacity of creating procedures that allow a business to react to the environment in an adaptive manner (Bower, 1974; Levinthal, 1997).

This leads to some considerations on aspects like entrepreneurship, innovation, and technological dynamics, which include questions like the dependence on history (*path dependence*), closure (*lock in*), adaptive learning, and the technological pathways in the area of resource economy. New knowledge resources are created from the moment that businesses discover new ways of undertaking their activity, and as a consequence other businesses learn from these improvements. One of the critical pathways for the creation of new knowledge resources comes through the development and standardization of new technologies.

If we think of the pharmaceuticals sector in which the methodology of *discovery* (discovery of a new drug) with respect to *rational drug design* has recently been complicated by the impact of two revolutionary technological matrices, microelectronics and biotechnology, on the industry. The knowledge required to manage biotechnologies has led to a scientific specialization of notable importance, requiring the full-time use of specialists who work with advanced instruments on macromolecules (essentially peptides or proteins), which are different by nature and dimension to the classical "objects" that were examined in classical chemical synthesis. As well as an increasing use of biotechnology techniques in the laboratory, entrepreneurial specialization that has germinated many initiatives has also been seen: the number of businesses, both public and private, specialized in the use of genetic engineering has notably increased, and the erosion of technological dominion to the detriment of many of the *incumbent* businesses has opened market space and partnership opportunities to an array of new organizations specializing in research, which have emerged from spinoff processes in the university setting or in R&D departments of large businesses, able to offer *incumbents* the capacity of innovation and complementary knowledge *assets*.

Traditional economic analysis does not apply to the standardization process, which is generally discussed only in modern literature on theories regarding technological dynamics.

In resource economy, standardization represents a central and critical process: it is the process through which a new resource, accessible to everyone, is created. The point of view of resources implies that, in the end, a perspective of the whole resource cycles in the economic world is taken into consideration, as occurs analogously in the biological world for the carbon and water cycles. In this instance, we shall not talk about resource cycles in a physical sense (that is, in terms of their material constituent elements) but rather in terms of creation, circulation, and destruction of the entity that generates value. A productive and satisfactory economy is certainly able to furnish a large variety or diversity of knowledge resources, which, in turn, imply processes of creation of resources and opportune disposal of resources no longer necessary. This creates what may be defined as a dynamic "equilibrium of resources," which can be explained in the sense in which the term is used in ecological analyses, which is essential for a healthy economy that generates a diversity of resources aimed at giving direction to adjustment and knowledge.

Markets for Knowledge Resources and the Generation of New Knowledge

In the setting of a real economy, resources are found in a state of constant mutation, thereby influencing the phenomenon of competitive and evolutionary dynamics. Businesses develop and exchange knowledge through procedures like free market operations, as in the case of the transfer of a division from one business to another, or more commonly, contractual agreements of various types (for example, those for the transfer of technology and the concession of licenses) or again through the transfer of resources following mergers and acquisitions.

Economists have been slow in recognizing the presence and importance of these multiple contacts between businesses within an industrial economy, recognizable in the figures of entrepreneurs, collaborators, suppliers, clients, and even competitors. And it is through these contacts that businesses exchange and share resources, voluntarily or involuntarily, through market and *non-market transactions*. These processes are identified as cases of diffusion, reproduction, exchange, redistribution, speculation (*leverage*), and sharing of knowledge resources; they belong to a strategic structure rather than to a deterministic cause–effect structure developed within the traditional economy and are involved in the dynamics of resource economy.

The supporting element of the functioning logic behind markets for knowledge resources, which also derive from the analyses made in the previous pages, lies in the fact that awareness, matured on the part of businesses, of the strategic value of sourcing maneuvers (*internal vs. external sourcing*) with respect to the needs posed by competition, constitutes an instrument that favors the creation of a context with knowledge bases suitable for perceiving the advantages of learning and integration and for elaborating the development of new knowledge (Argote and Miron-Spektor, 2011; Leiponen and Helfat, 2010; Garriga et al. 2013; Chiang and Hung, 2010).

Therefore, the links between entrepreneurial behavior and mechanisms of generating knowledge may be read in two different keys.

In the first instance, while on the one hand, the efficiency of the solutions of division of labor and vertical disintegration, to use the most common language of economy, has a strong influence on the entry conditions of new businesses and new types of business, the competitive position of existing businesses and the structure of the sector, on the other hand, obtain a notable increase in the capacity for the development of innovation, with a better use of knowledge, especially in a wider sense. In the second instance, looking from the internal perspective, the efficiency of businesses in the management and use of knowledge resources, as an effect of the existence of *markets for knowledge resources* and the variety within the economic system, exalts the role played by the endowment of resources that determine the diversity of performance between businesses, development pathways of new knowledge, and their future boundaries; considering projection toward the outside, the evolution of a business involves the environmental system and produces a form of co-evolution, which hypothesizes the functioning of a business based on routines of knowledge, which develop and are selected according to their capacities to adapt, which influence the main strategic, organizational, and governance decisions and represent the determining principle of competitive advantage.

Chapter Summary

While it is certain that the choice of exploring the theme of markets for knowledge resources initiates from the need for a better understanding of how pragmatic alternatives affect the innovative capacity of a businesses, our work also takes inspiration from the new theoretical developments that identify the existence of business in a co-evolutionary relationship with the environment, through knowledge found beyond the confines of business and the need of accumulating it and using it efficiently.

NOTES

1. In Great Britain, for example, between 1985 and 1995 alone, there was a doubling, in real terms, of the total figures for the phenomenon. From 1985 to 1995 the figures passed from £450 million to £935 million. This surprising data does not concern only the total growth of BERD (Business Expenditure of Research and Development), which in this timespan grew only 14 percent, but also that of external contracting which, with respect to the BERD total, passed from 5.5 percent (in 1985) to 10 percent (Source: *Office of National Statistic data*). Whittington, R., "The Changing Structure of R&D: From Centralization to Fragmentation," in: R. Loveridge and M. Pitt (Eds), *The Strategic Management of Technological Innovation*. London: Wiley, pp. 183–203, 1990.
2. In particular, for example, the use of consulting engineers has delayed innovative process and organizational evolution in the electrical energy sector in Britain (Byatt, I.C.R., *The British Electrical Industry, 1875–1914.* Oxford: Oxford University Press, 1979).
3. Even in sectors with a long scientific tradition, like the pharmaceutical industry, up until the First World War there was a strong stimulus toward "opening up to the outside," to feed the research and development processes, at the same rate as the growing scientification of technologies and the reduction of times which, in some fields, characterized the passage from scientific discovery to industrial applications (Liebenau, J. M., "International R&D in Pharmaceutical Firms in the Early Twentieth Century," *Business History*, 26, 1984, 329–346; Swan, P., *Academic Scientist and the Pharmaceutical Industry: Cooperative Research in Twenty-Century America*. Boston, MA: John Hopkins University Press, 1989).
4. In many countries like Great Britain, the emergence and later development of internal R&D departments arrived later, and it was certainly less widespread and significant compared to the USA, and even less so than in Germany where policies for the promotion and support of industrially-based research activities of the universities and public centers gave useful results for technological applications, consolidating the development of industry-academic links (Mowery, D. C., "Firm Structure, Government Policy, and The Organization of Industrial Research: Great Britain and the United States, 1900–1950," *Business History Review*, 58, 1980, 504–531; Mowery, D. C. and Rosenberg, N. *Technology and the Pursuit of Economic Growth.* Cambridge: Cambridge University, 1989; Meyer-Thurow, "The Industrialization of Inventino: A Case Study from the German Chemical Industry," *Isis*, 73, 1982, 363–381).
5. Lanza, A., *Knowledge Governance. Dinamiche Competitive e Cooperative nell'Economia delle Conoscenza.* E.G.E.A.: Milano, 2000.

6. On this question see: Winter, S., "Knowledge and Competence As Strategic Asset," in: D. Teece (Ed.), *The Competitive Challenge: Strategies for Industrial Innovation and Renewal*, Cambridge, MA: Ballinger, pp. 159–183, 1987. Prahalad, C. and Hamel, G., "The Core Competence of Corporation," *Harvard Business Review*, 35, 1994, 43–55.
7. Coombs, R., "Core Competencies and the Strategic Management of R&D," *R&D Management*, 26, 1996, 345–355.
8. Cohen, W. M. and Levinthal, D. A., "Absorptive Capacity: A New Perspective on Learning and Innovation," *Administrative Science Quarterly*, 35, 1, 1990, 128–152.
9. Another argument can be made to support the notion that internal and external R&D are complementary and do not substitute each other. Discussing the complex relationship between basic and applied research, Rosenberg stresses that a high level of R&D may be useful in solving problems which arise during the production phase. Rosenberg's point of view is that a business which lacks the technical capacities will often be unable to define the problems it faces in an adequate manner, through systematic scientific investigation. In a similar way, von Hippel showed that a lot of innovations come to the very clients that try to resolve problems that occur when they extend their technological frontiers. *Independent technology suppliers* have no incentive to supply solutions to these problems, at least not before those who develop the technology have to face similar problems. After all, the development of new knowledge depends on chance, and the solutions foreseen for one context may be used in others.
10. Reflecting on Penrose's claim:

    A theory of business development is essentially an examination of the changes in production opportunities for businesses; if one wishes to find a limit to the development of businesses, or a conditioning of the rate of development, one needs to take into account that production opportunities are limited in every period. It is clear that these opportunities are reduced in proportion to the capacity of a business to predict expansion opportunities, to the willingness or not to exploit them, and to the ability of knowing how to exploit them opportunely. (1959)

11. During the contract negotiations no attempt is made to give a definition to the technology, rather it is illustrated with a vague term indicating some useful knowledge which has its roots in the field of engineering or the scientific disciplines and usually also draws on the practical experience of production. Transactions in the field of technology can have differing forms, from the simple concession of a licence to well-defined intellectual property, to complicated collaboration agreements which may also include a further development of the technology or an "ex novo" creation of the same.

12. The division of innovative labor is characterized by the interdependences between the various stages or activities involved in the innovative process for which a partitioning of success in innovative labor requires an innovative process to be effectively decomposable. The work of Herbert Simon explains that the decomposition of a complex problem into more elementary problems becomes an organizational question. The level at which one may plan a complex activity of problem solving, so as to reduce the interdependency between tasks, certainly influences, in turn, the capacity of accurate definition of the various phases. For example, the artisan of the pre-capitalist era, who carried out all the activities regarding the conceptualization and manufacture of his products, found it difficult, mentally and functionally, to separate his activity into independent operations, while the Fordists system of production (with Adam Smith's pin industry) was able to effectively identify the criteria and methods of partitioning.
13. Teece, D. J., "Capturing Value from Knowledge Assets: The New Economy, Markets for Know-how, and Intangible Assets," *California Management Review*, 40, 3, 1998, 55–79.
14. Hart, O. and Moore, J., "Property Rights and the Nature of the Firm," *Journal of Political Economy*, 98, 1990, 1119–1158.
15. Arrow, K. J., *The Limits of Organization*, New York: W.W. Norton, 1974.
16. The problem of "theft" (= the "hold-up" problem), shown in the literature regarding transaction costs (v., ad es., Williamson, 1985; Klein et al., 1978), comes about when one party is able to expropriate income from another.
17. In the case, then, that the level of the "TC" tax is low and the profits deriving from the exchange even lower, there will be no clear motive to justify vertical specialization because of the symmetrically distributed capacities. In the same way, if the increase in transaction costs is not sufficient to exceed the profits of specialization, integration shall be the option taken, A decrease in the transaction costs facilitates market mechanisms because it allows businesses to capitalize on their capabilities and on their relative points of strength. If businesses are different, a reduction of transaction costs will allow a significant specialization, in that each business focalizes on its own area of strength. If, instead, all the businesses differ in equal proportions in the various stages, the same reduction will not achieve promotion of specialization and disintegration.
18. The so-called *arm's-length contracts*, in the exchange of (tacit) knowledge resources, for example, are characterized by problems of hypocrisy and *moral hazard*. Arora, A. and Merges, R. P., "Specialized Supply Firms, Property Rights and Firm Boundaries," *Industrial and Corporate Change*, 3, 3, 2004, 451–475.
19. Freeman, C. and Soete, L., *The Economics of Industrial Innovation*. London: Pinter, 1997.

20. The members of Group S, constituted as the separate Business S, could create better results.
21. Masten, S. E., *Case Studies in Contracting and Organization*. New York: Oxford University Press, 1996.
22. Williamson, O. E., *The Mechanisms of Governance*. New York: Oxford University Press, 1996.
23. There are contractual limits. Ex ante, it could be difficult or impossible to estimate the final price of S for the suitable input. It could be impossible to draft an executive contract to describe the level of effort of S, employed in adapting its input to the needs of B.
24. Implicit in the approach of Arora and Merges is the conviction that in some cases *specialized suppliers* possess potential for generating useful know-how and building capacity at the level of the business, greater than that which would be had in a large integrated business. That creates incentives for building up know-how and complementary assets.
25. The definition of "technology market" is adapted by Arora, Fosfuri, Gambardella (2001) who take inspiration from that proposed by the American Department of Justice contained in the "Anti-trust Regulation on the Concession of Licenses of Intellectual Property" (United States Department of Justice, 1995). The American Department of Justice defines the technology markets as markets of intellectual properties conceded in licenses and their substitute products, that is to say those technologies and other instruments that can substitute them in such an incisive way as to impose the exercise of the power of the market with respect to the intellectual property conceded by licensing (United States Department of Justice, 1995).
26. On the discussion of the primary role of IPRs, Arora and Merges have built their contributions both on positions regarding "capabilities" (Teece et al.), and on that of "property rights" (Hart). Arora, and Merges, "Specialized Supply Firms, Property Rights and Firm Boundaries"; Teece, D. J., Pisano, G. and Shuen, A., "Dynamic Capabilities and Strategic Management," *Strategic Management Journal*, 18, 1999, 509–533. Hart, O., *Firms, Contracts, and Financial Structures*. Oxford: Oxford University Press, 1995.
27. Pisano, G., Shan, W. and Teece, D., "Joint Ventures and Collaboration in Biotechnology Industry," in: D. Mowery (Ed.), *Technological Collaboration Ventures in US Manufacturing*, Cambridge, MA: Ballinger, pp. 183–222, 1988.
28. On this question the work presented by Cesaroni is interesting. Its scope was to estimate the possibility of producing (and not commercializing) technologies, in an open market, and draws on business decisions on the research of technological sources, and their propensity to enlarge their range thanks to the entrance of new products into the markets. The study carried out by Cesaroni focalizes on the use

of information regarding production plants of 96 chemical businesses in the United States and Japan. This has shown that the producers of chemicals have created some strong exchange relationships in R&D with engineering companies—specialized engineering firms (SEFs)—situated upstream in the production chain, and specializing in the planning and engineering of chemical processes. SEFs have some important implications on the structure of the market. In a particular way, SEFs were an important source of technology and know-how in the planning of plants, which allowed helping new businesses to enter the chemicals market. Usually SEFs offer a complete technology package, consistent in the core technologies (often given in licensing by an incumbent chemicals producer), with installation and engineering services. As a consequence, technological outsourcing and the entry in new segments of the sector were facilitated by the presence of SEFs and by the relative market for technological processes. Certainly the division of labor remains a complex process, which requires dense interactions between the two businesses, and which can only be carried out if the chemical company maintains its key competitiveity intra moena. For example one of these "*core competencies*," for the business can consider itself satisfied, is the capacity of transferring (or applying) the business opportunities in the required technologies. In the same way, the capacity regarding the critical components of the plants must be requalified within the chemicals industry. Hence, technological, organizational and strategic considerations come into play during the process. Therefore, the presence of SEFs creates a lot of incentives to substitute the internal technological development with external outsourcing. Cesaroni, reaches two hypotheses, the first is that "in the presence of a larger technology market, businesses tend to substitute internal R&D investments or long term contracts with contracts (of reciprocal independence) of short-period technological outsourcing." The second is that "in the presence of a larger technology market, businesses tend to diversify more." Cesaroni, F., "Technological Outsourcing and Product Diversification: Do Markets for Technology Affect Firms' Strategies?" *Research Policy*, 33, October 2004, 1547–1564.

29. There are some ways through which the efficiency of contracts can increase. These regard the construction of *reputation*, when we are confronted with a repeat contract, and the use of output-based *royalties*. In fact, output-based *royalties* may not resolve the problem of *moral hazard*, in that the total output produced by the *licensee* is often a private piece of information and difficult to estimate for the *licensor* or others. Output-based *royalties* could, besides, impede the *licensee* in the product market, especially in oligopolist markets (Katz and Shapiro, 1985). For this reason the compensation the *licensor*'s technical assistance through the use of output-based royalties is

not very common (Contractor, 1981). The building of a reputation through forms of repeated contract, despite being a potential solution, provides a higher level of integration between partners. Normally markets imply anonymous transactions and the difficulty lies in verifying whether technology transactions can be increased even without establishing a reputation and long-term relationship between the *contractors*. Efficiency of contracts for the exchange of technology can be reached through the exploitation of the complementary nature of know-how and other technological inputs which the concessionary can use as a "guarantee." Through its complementary nature, know-how becomes precious when it is used in conjunction with complementary technological instruments which may be withdrawn. This allows the *licensor* to make use of its capacities and to withdraw, thereby creating a shield against possible opportunistic behavior on the part of the *licensee*.

The concessionary protects itself postponing part of the payment to the time of complete liberation of know-how. If the concessionary does not carry out the second payment, the concedant can withdraw from the contract and therefore withdraw the inputs of a complementary nature. While the additional benefits of the transfer of know-how and complementary inputs on the part of the concedant are greater than the costs supported by the concedant for the offer of know-how and complementary inputs, the concedant will honor the contract as best possible. Thus, the problem of opportunism can be mitigated through simple, self-regulating contracts. Patents and other types of IP can work better thanks to the complementary input produced by the concedant. A typical case can be that in which the technology to be transferred is composed of both patentable components and complementary know-how (experience with the use of technology). In these cases, the concedant can withdraw the patent rights the moment the execution of the contract is not satisfied. With strong and well-defined patents, the concessionary cannot but consider that the value is attributable to a great extent, and in an isolated way, to know-how.

30. The transaction cost economics theory shows the fundamental concept that simple market contracts do not adequately protect against risks of expropriation in cases where investments in specific resources are involved (that is to say that the resources cannot be transferred to a better use or to a successive user, without this leading to a significant loss of value) (Klein et al., 1978). The theory of transaction costs considers (1) that businesses internalize transactions where highly specific activities are involved ("produce" instead of "buy"), or (2) that they underinvest in areas where there are high risks of expropriation (Williamson, 1985).
31. This intuition is particularly incisive in a context of badly defined activity, as for example, intellectual property. As previously observed,

establishing up to what point the patent rights requested by one player may violate the patent rights already obtained by other players is very difficult and very costly (Merges and Nelson, 1990). Besides, since the economic value of the intellectual property depends on the use of the same within a particular competitive or technological setting, it is regarded as highly context specific (Teece, 1986). These characteristics of patent rights and of exchange processes indicate that the contractual costs connected to IP depend, in a crucial way, on how external rights are distributed.

32. Jonash (1996) refers to the technological aspect of the process as to "*strategic technology leverage*" so that businesses take up the challenge for the governance of their portfolios of complex technology, while Wolff (1992) uses the phrase "*technology adoption*" to represent the concept of an assimilation connected to *external sourcing*.
33. Dierickx and Cool (1989) consider the competitive advantages which are believed to characterize a business which chooses exclusively internal development of resources. the process of accumulation of resources takes advantage of *time compression diseconomies*, of *asset mass efficiencies*, of *asset stock interconnectedness*, of the *prevention of asset erosion* and of *causal ambiguity*. In the style of these authors, we may qualify the potential advantages for businesses which adopt the choice of external sourcing of resources as those deriving from the accelerated speculation of resources, from their variety and from resources in *free-riding*, from their complementary nature and novelty and finally from knowledge factors like *causal clarity*, as in the case of modularization of knowledge and the change associated with it which transforms from implicit knowledge to explicit knowledge.

REFERENCES

Argote, L, & Miron-Spektor, E. Organizational learning: from experience to knowledge. *Organization Science*, 22: 1123–1137 (2011).

Arora, A., Fosfuri, A., & Gambardella, A. *Market for Technology: The Economics of Innovation and Corporate Strategy*. Cambridge, MA: MIT Press (2001).

Arora, A., & Merges, R. P. Specialized supply firms, property rights and firm boundaries. *Industrial and Corporate Change*, 3(3) (2004).

Arrow, K. J. *The Limits of Organization*. New York: W.W. Norton (1974).

Barley, R. S., & Tolbert S. P. Institutionalization and structuration: Studying the links between action and institution. *Organization Studies*, 18(1) (1997).

Barney, J. B. Strategic factor markets: Expectations, luck and business strategy. *Management Science*, 32 (1986).

Barney, J. Firm resources and sustained competitive advantage. *Journal of Management*, 17(1) (1991).

Barney, J. Looking inside for competitive advantage. *Academy of Management Executive*, 9(4) (1995).

Berger, P., & Luckmann T. *The Social Construction of Reality: A Treatise in the Sociology of Knowledge*. Peregrine Books (1979).

Bitner, M. J. Evaluating service encounters: The effect of physical surroundings and employee responses. *Journal of Marketing*, April (1990).

Bollen, K. A. *Structural Equations with Latent Variables.* New York: Wiley (1989).

Bower, J. L. Planning and control: Bottom up or top down?. *Journal of General Management*, 3:22–23 (1974).

Brown, J. S., & Duguid, P. Organizing knowledge. *California Management Review*, Spring, 4(3) (1988).

Brusoni, S., & Prencipe, A. Unpacking the black box of modularity: Technologies products and organizations. *Industrial and Corporate Change*, 10 (2001).

Burgelman, R. A. Fading memories: a process theory of strategic business exit in dynamic environments. *Administrative Science Quarterly*, 39(1): 24–56 (1994).

Burns, J., & Scapens, R. W. Conceptualising management accounting change: An institutional framework. *Management Accounting Research*, 11 (2000).

Byatt, I. C. R. *The British Electrical Industry, 1875–1914*, Oxford: Oxford University Press (1979).

Cesaroni, F. Technological outsourcing and product diversification: Do markets for technology affect firms'strategies? *Research Policy*, 33(10), 1547–1564 (2004).

Chiang Y- &, Hung K-P. Exploring open search strategies and perceived innovation performance from the perspective of inter-organizational knowledge flows. *R&D Management*, 40(3): 292–299 (2010).

Choo, F. Auditors' knowledge and judgement performance: A cognitive script approach. *Accounting, Organizations and Society*, 21(4) (1996).

Christensen, C. M., & Rosenbloom, R. S. Explaining the attacker's advantage: Technological paradigms, organizational dynamics and the value network. *Research Policy*, 24 (1995).

Churchill, G. A. Jr. A paradigm for developing better measures of marketing constructs. *Journal of Marketing Research*, 16, February (1979).

Coase, R. H. The institutional structure of production. *The American Economic Review*, 82(4) (1992).

Coase, R. H. The problem of social cost, in idem. 1988. *The Firm, the Market and the Law*, Chicago: University of Chicago Press (1960).

Cohen, M. D., & Bacdayan, P. Organizational routines are stored as procedural memory: Evidence from a laboratory study. *Organization Science*, 5(4) (1994).

Cohen, M. D., Burkhart, R., Dosi, G., Egigi, M., Marengo, L., Warglien, M., & Winter, S. Routines and other recurring action patterns of organizations: Contemporary research issues. *Industrial and Corporate Change*, 5 (1996).

Cohen, W. M., & Levinthal, D. A. Absorptive capacity: A new perspective on learning and innovation. *Administrative Science Quarterly*, (35) (1990).

Cohendet, P., & Llerena, P. Routines and incentives: The role of communities in the firm, *Industrial and Corporate Change*, 12(2) (2003).

Colombo, M., & Gerrone, P. Technological co-operative agreements and firm's R&D Intensity. *Research Policy*, 25 (1996).

Constant, & Edward. The social locus of technological practice: Community, system, or organization?, In Bijker, Hughes, & Pinch (Eds.) *The Social Construction of Technological Systems* (Cambridge, MA: MIT Press) (1999).

Coombs, R. Core competencies and the strategic management of R&D. *R&D Management*, 26 (1996).

Cossette, P., & Audet, M. Mapping of an idiosyncratic schema. *Journal of Management Studies*, 29/3 (1992).

Cowan, R., David, P. A., & Foray, D. The explicit economics of knowledge codification and tacitness. *Industrial and Corporate Change*, 2 (2000).

Davenport, T. H., & Prusak, L. *Il sapere al lavoro.* ETAS (2000).

Davis, Lance, & Douglass, C. *North, Institutional Change and American Economic Growth.* Cambridge: Cambridge University Press (1971).

Dell'Anno, D., & Del Giudice, M. Il ruolo della conoscenza nei fenomeni di gemmazione d'impresa: un'evidenza empirica da contesti accademici, in Brondoni, S. M. (Ed.) *Il sistema delle risorse immateriali d'impresa: cultura d'impresa, sistema informativo e patrimonio di marca.* Giappichelli, Torino (2004).

Dell'Anno, D., & Del Giudice, M. The spin off model: a simultaneous way to knowledge transfer and entrepreneurship stimulation. *In 7th World Congress on Total Quality Management*, CUEIM (2002).

Dell'anno, D., Van der Sijde, P., & Del Giudice, M. A script-based approach to spin off: Some first issues on innovative pathways of knowledge transfer and academic knowledge's reproducibility, In Oakey, R., During, W., & Kauser S. (Eds.) *New Technology-based Firms in the New Millenniu.* Elsevier (2007).

Demsetz, H. *The Organization of Economic Activity.* New York: Blackwell (1988).

Demsetz, M. Towards a theory of property rights. *American Economic Review* (1967).

Doz, Y., & Santos, J., & Williamson, P. *From Global to Metanational: How Companies Win in the Knowledge Economy.* Boston, MA: Harvard Business School Press (2001).

Drucker, P. F. *Post-capitalist Society.* New York: Harper Business (1993).

Eden, C. On the nature of cognitive maps. *Journal of Management Studies*, 29/3 (1992).

Edwards, D. Script formulations: An analysis of event descriptions in conversation. *Journal of Language and Social Psychology*, 13(3) (1994).

Eisenhardt, K. M. Building theories from case study research. *Academy of Management Review*, 14(4) (1989).

Evrard, Y., Pras, B., & Roux, E. *Market, etudes et recherches en marketing.* Ed. Nathan, Paris (1993).

Foray, D., & Steinmueller, W. E. The economics of knowledge reproduction by inscription. *Industrial and Corporate Change*, 2 (2003).

Fornell, C. R., & Lacker, D. F. Two structural equation models with unobservable variables and measurement error. *Journal of Marketing Research*, 18 (1981).

Foss, N. J., & Klein, P. G. *Organizing Entrepreneurial Judgment*. Cambridge University Press: Cambridge, MA (2012).

Freeman, C., & Soete, L. *The Economics of Industrial Innovation*. London: Pinter (1997).

Gambardella, A. Competitive advantages from In-house basic research. *Research Policy*, 21 (1992).

Garriga, H., von Krogh, G., & Spaeth, S. How constraints and knowledge impact open innovation. *Strategic Management Journal*, 34: 1134–1144 (2013).

Garud, R., & Rappa, M. A socio-cognitive model of technology evolution: The case of cochlear implants. *Organization Science*, 5(3): (1994).

Gawer, A., & Cusumano, M. A. *Platform Leadership: How Intel, Microsoft, and Cisco Drive Industry Innovation*. Harvard Business School Press (2002).

Gerbing, D. W., & Anderson, J. C. An updated paradigm for scale development incorporating unidimensionality and its assessment. *Journal of Marketing Research*, 25, May (1988).

Giddens, A. *The Constitution of Society*. Berkeley, CA: University of California Press (1984).

Gioia, D., & Poole, P. Scripts in organizational behavior. *Academy Management Review*, 9 (1984).

Glaser, B., & Strauss, A. *The Discovery of Grounded Theory*. Chicago: Aldine (1967).

Grant, R. The resource based theory of competitive advantage: Implications for strategy formulation. *California Management Review*, Spring (1991).

Grant, R. Impresa e organizzazione. *Sviluppo & Organizzazione*, 169 (1998).

Guida, G., & Berini, G. *Ingegneria della conoscenza*. EGEA (2000).

Hamel, G. Competition for competence and inter-partner learning within international strategic alliances. *Strategic Management Journal*, Winter Special Issue, 12 (1991).

Hannan, M., & Freeman, J. *Ecologia organizzativa Per una teoria evoluzionista dell'organizzazione*, Etaslibri, Milano (1993).

Hart, O. *Firms, Contracts, and Financial Structures*. Oxford: Oxford University Press. (1995).

Hart, O., & Moore, J. Property rights and the nature of the firm. *Journal of Political Economy*, 98 (1990).

Heeler, R., & Ray, M. Measure validation in marketing. *Journal of Marketing Research*, 9, November (1972).

Heller, M. A. The boundaries of private property. *Yale Law Journal* (1999).

Heller, M. A., & Eisenberg, R. S. Can patents deter innovation? The Anticommons in biomedical research. *Science*, 280 (1998).

Henderson, J., & Leleux, B. Corporate venture capital: Effecting resource combinations and transfers. *Babson Entrepreneurial Review*, May (2002).

Henderson, R. M., & Clark, K. B. Architectural innovation: The reconfiguration of existing systems and the failure of established firms. *Administrative Science Quarterly*, 35 (1990).

Hill, C. W. L., & Rothaermel, F. T. The performance of incumbent firms in the face of radical technological innovation. *Academy of Management Review*, 28 (2003).

Huang, K.-T. Capitalizing on intellectual assets. *IBM Systems Journal*, 4 (1998).

Ireland, R. D., Covin, J. G., & Kuratko, D. F.. Conceptualizing corporate entrepreneurship strategy. *Entrepreneurship Theory and Practice*, 33: 19–46 (2009).

Jacobides, M. G. How capabilities, transaction costs and scalability interact to drive vertical scope, *Working Paper*, London Business School (2004).

Jacobides, M. G., & Hitt, L. M. Losing sight of the forest for the trees? Productive capability differences as drivers of vertical scope, *Leverehulme Working Paper*, London Business School (2004).

Jacobides, M. G., & Winter, S. G. The co-evolution of capabilities and transaction costs: explaining the institutional structure of production. *Strategic Management of Journal*, 26 (2005).

Jensen, M. C., & Meckling, W. H. Theory of the firm, managerial behavior, agency costs and ownership structure. *Journal of Financial Economics*, 3(4) (1976).

Johannessen, J.-A., Olaisen, J., & Olsen, B. Mismanagement of tacit knowledge: The importance of tacit knowledge, the danger of information technology, and what to do about it. *International Journal of Information Management*, 21 (2001).

Johnson-Laird, P. *The Computer at the Mind*. Cambridge: Harvard University Press (1988).

Jonash, Ronald S. Strategic technology leveraging: Making outsourcing work for you. *Research-Technology Management*, March–April (1996).

Jong, S. How organizational structures in science shape spin off firms; The biochemistry departments of Berkeley, Stanford, and UCSF and the birth of the biotech industry. *Industrial and Corporate Change*, 2 (2006).

Katz, M., & Shapiro, C. On the licensing of innovation. *RAND Journal of Economics*, 16(4) (1985).

Kim, D. H. The link between individual and organizational learning. *Sloane Management Review*, Fall (1993).

Klein, B., Crawford, R., & Alchian, A. Vertical integration, appropriable rents, and the competitive contracting process. *Journal of Law and Economics*, 21

Klein, B., Crawford, R. G., & Alchian, A. A. Vertical Integration, Appropriable Rents, and the Competitive Contracting Process, *Journal of Law and Economics*, 21: 297–326 (1978).

Kleinknecht, A., & van Reijnen, J. Why do firms cooperate in R&D: An empirical study. *Research Policy*, 21 (1992).

Knorr Cetina, K. *Epistemic Cultures: How the Sciences Make Knowledge*. Cambridge: Harvard University Press (1999).

Kogut, B., & Zander, U. Knowledge of the firm, combinative capabilities, and the replication of technology. *Organization Science*, 3(3) (1992).

Kolb, D. *Experential Learning*. New Jersey: Prentice Hall (1984).

Krieger, F., & Stratmann, S. *The Universitat Dortmund: An Agent of Structural Change in its Region*. Twente University Press (2000).

Langlois, R. N. Modularity in technology and organization. *Journal of Economic Behavior and Organization*, 49 (2002).

Langlois, R. N, & Foss, N. J. Capabilities and governance: The rebirth of production in the theory of economic production. *Kyklos*, 52(2) (1999).

Langlois, R. N., & Robertson, P. L. *Firms. Markets and Economic Change: A Dynamic Theory of Business Institutions*. London: Routledge (1995).

Lanza, A. Knowledge Governance. *Dinamiche Competitive e Cooperative nell'Economia delle Conoscenza*. Milano: E.G.E.A. (2000).

Leiponen, A., & Helfat, C. E. Innovation objectives, knowledge sources, and the benefits of breadth. *Strategic Management Journal*, 31: 224–236 (2010).

Leonard Barton, D. Core capabilities and core rigidities. A paradox in managing new product development. *Strategic Management Journal*, 13 (1992).

Levin, R. C., Klevorick, A. K., Nelson, R. R., & Winter, S. G. Appropriating the returns from industrial research and development. *Brookings Papers on Economic Activity*, 3 (1987).

Levinthal, D. Adaptation on rugged landscapes. *Management Science*, 43(7) (1997): 934–950.

Libecap, G. *Contracting for Property Rights*. Cambridge: Cambridge University Press (1989).

Liebenau, J. M. International R&D in pharmaceutical firms in the early twentieth century. *Business History*, 26 (1984).

Lieberman, M. B., & Montgomery, D. First mover advantages. *Strategic Management Journal*, 9 (1988).

Lippman, S., & Rumelt, R., Uncertain imitability: An analysis of interfirm differences in efficiency under competition. *The Bell Journal of Economics*, 13 (Autumn) (1982).

Louis, M. R. Surprise and sense making: What newcomers experience in entering unfamiliar organizational settings. *Administrative Science Quarterly*, 25 (1980).

Maggioni, V. *L'impresa come sistema socio-tecnico di tipo aperto*. Napoli (1981).

Maggioni, V. Apprendere dalle strategie relazionali delle imprese: Modelli ed esperienze per le meta-organizzazioni. *Sinergie*, 52 (2000).

Maggioni, V., & Del Giudice, M. Relazioni sub-sistemiche, diffusione d'imprenditorialità interna e processi di gemmazione: una verifica empirica. In Brondoni, S. M. (Ed.) *Il sistema delle risorse immateriali d'impresa: cultura d'impresa, sistema informativo e patrimonio di marca, Giappichelli*, Torino (2004).

Maggioni, V., & Del Giudice, M. Il ruolo della script analisys nei processi di creazione d'impresa: uno studio empirico su imprenditorialità diffusa e gemmazioni, In Amenta P., D'Ambra, L., Squillante, M., Ventre, A., & Metodi. (Eds.) *Modelli e Tecnologie dell'informazione a supporto delle decisioni*, Franco Angeli (2005).

Maggioni, V., & Del Giudice, M. Relazioni sistemiche tra imprenditorialità interna e gemmazione d'impresa: una ricerca empirica sulla natura cognitiva delle nuove imprese. *Sinergie*, 1 (2006)

Mahoney, J. T., & Pandian, J. R. The resource-based view within the conversation of strategic management. *Strategic Management Journal*, 13 (1992).

Malerba, F., & Orsenigo, L. Technological regimes and sectoral patterns of innovative activities. *Industrial and Corporate Change*, 6 (1997).

March, J. G., & Simon, H. A. *Organizations.* New York: John Wiley and Sons (1958).

Markides, C. C., & Williamson, P. J. Corporate Diversification and Organizational Structure: A Resource-Based View. *Academy of Management Journal*, 39 (1996) 340–367.

Masten, S. E. *Case Studies in Contracting and Organization.* New York: Oxford University Press (1996).

Mathews, J. A. Strategizing by firms in the presence of markets for resources. *Industrial and Corporate Change*, 12(6) (2003).

Merges, R. P., & Nelson, R. R. On the complex economics of patent scope. *Columbia Law Review*, 90(4) (1990).

Meyer-Thurow. The industrialization of inventino: A case study from the German chemical industry. *Isis*, 73 (1982).

Mintzberg, Henry. The strategy concept I: Five Ps for strategy. *Management Review* Fall (1987): 11–16.

Mol, M., & Birkinshaw, J. M. The sources of management innovation: when firms introduce new management practices. *Journal of Business Research*, 62 (2009): 1269–1280.

Mowery, D. C. Firm structure, government policy, and the organization of industrial research: Great Britain and the United States, 1900–1950. *Business History Review*, 58 (1980).

Mowery, D. C., & Rosenberg, N. *Technology and the Pursuit of Economic Growth.* Cambridge: Cambridge University (1989).

Nasbeth, L., & Ray, G. F. *The Diffusion of New Industrial Process.* Cambridge University Press (1974).

Nelson, S. G., & Winter, R. R. *An Evolutionary Theory of Economic Change.* Cambridge, MA: Harvard University Press (1982).

Nightingale, P. If Nelson and Winter are only half right about tacit knowledge, which half? A Searlean critique of codification. *Industrial and Corporate Change*, 2 (2003).

Nonaka, I., & Konno, N. The concept of "Ba": Building a foundation for knowledge creation. *California Management Review*, 40/3 (1998).

Nonaka, I., & Takeuchi, H. *The Knowledge Creating Company*. Oxford: Oxford University Press(1997).

Nonaka, I., Toyama, R., & Byosieère, P. A theory of organizational knowledge creation: Understanding the dynamic process of creating knowledge. In Dierkes, M., Berthoin Antal, A., Child, J., & Nonaka, I. (Eds.) *Handbook of Organizational Learning & Knowledge*, Oxford University Press (2001).

Nonaka, I., Toyama, R., & Konno N. SECI, Ba and leadership: A unified model of dynamic knowledge creation. *Long Range Planning*, 33 (2000).

Nooteboom, B. *Learning and Innovation in Organizations and Economies*. New York: Oxford University Press (2000).

North, Douglass C., & Thomas, Robert P. *The Rise of the Western World: A New Economic History*. Cambridge, UK: Cambridge University Press (1973).

Nunnally, J. C., & Bernstein, I. H. *Psychometric Theory*. New York, NJ: McGraw-Hill (1994).

Penrose, E. T. *La teoria dell'espansione dell'impresa, Franco Angeli*. Milano (1973) (ed. orig. 1959).

Penrose, E. *The Theory of Growth of the Firm, Blackwell*. New York: Oxford & John Wiley & Sons (1959).

Pérez-Luño, A., Wiklund, J., & Cabrera, R. V. The dual nature of innovative activity: how entrepreneurial orientation influences innovation generation and adoption. *Journal of Business Venturing*, 26: 1–17 (2011).

Peteraf, M. A. The cornerstones of competitive advanatge: A resource-based view. *Strategic Management Journal*, 14 (1993).

Pierce, J. R., & Aguinis, H. The too-much-of-a-good-thing effect in management. *Journal of Management*, 39: 313–338 (2013).

Pisano, G. The governance of innovation: Vertical integration and collaborative arrangements in the biotechnology industry. *Research Policy*, 20 (1991).

Pisano, G., Shan, W., & Teece, D. Joint ventures and collaboration in biotechnology industry, In Mowery, D. (Ed.) *Technological Collaboration Ventures in US Manufacturing*. Cambridge, MA: Ballinger (1988).

Polany, M. The Tacit Dimension (1966), In Prusak, L. (Ed.) *Knowledge in Organizations*, Butterworth-Heinemann (1997).

Polanyi, M. *The Tacit Dimension*. London: Routledge (1967).

Porac, J. F. et al., Causal attributions, affect, and expectations for a day's work performance. *Academy of Management Journal*, 26(2) (1983).

Prahalad, C. K., & Hamel, G. The core competence of the corporation. *Harvard Business Review*, May–June (1990).

Prahalad, C., & Hamel, G. The core competence of corporation. *Harvard Business Review*, 35 (1994).

Quinn, J. B. Strategic change: "Logical Incrementalism". *Sloan Management Review*, 20(1): 7 (1978).

Quinn, J. B. *Intelligent Enterprise: A Knowledge and Service based Paradigm for Industry*. New York: The Free Press (1992).

Reed, R., & De Filippi, R. Causal ambiguity, barriers to imitation and sustainable competitive advantage. *Academy of Management Review*, 15(1) (1990).

Riordan, M. H., & Williamson, O. E. Asset specificity and economic organization. *International Journal of Industrial Organization*, 3 (1985).

Roberts, E. B., & Malone, D. E. Policies and structures for spinning off new companies from research and development organizations. *R&D Management*, 26(1) (1996).

Rogers, E. M. *The Diffusion of Innovation.* Glencoe, IL: The Free Press (1982).

Rosenberg, N. Why do firms do basic research. *Research Policy*, 19 (1990).

Sanchez, R., & Mahoney, J. Modularity, flexibility, and knowledge management in product and organization design. *Strategic Management Journal*, 17 (Winter Special Issue) (1996).

Schank, R. C., & Abelson, R. P. *Scripts, Plans, Goals and Understanding: An Inquiry into Human Knowledge Structures.* Hillsdale, NJ: Lawrence Erlbaum (1977).

Scharmer, C. O. Organizing around not-yet-embobied knowledge. In Von Krogh, G., Nonaka, I., & Nishiguchi, T. (Eds.) *Knowledge Creation, A Source of Value*, Palgrave (2000).

Schutte, F., & Van der Sijde, P. C. *The ECIU University.* University of Twente Press (2000).

Sciarelli, S. *Economia e Gestione dell'impresa.* Cedam: Padova (2002).

Sciarelli, S. *Fondamenti di Economia e Gestione delle Imprese.* Cedam: Padova (2004).

Sciarelli, S. *Il governo dell'impresa in una società complessa: la ricerca di un equilibrio tra economia ed etica, in Sinergie*, Gennaio-Aprile (1998).

Shapiro, C. Navigating the patent thickets: Cross-licenses, patent pools, and standard-setting. In Jaffe, A., Lerner, J., & Stern, S. (Eds.) *Innovation Policy and the Economy*, v1. Cambridge MA: NBER (2001).

Sharp, D. J. The effectiveness of routine-based decision processes: The case of international pricing. *Journal of Socio-Economics*, 23(1/2) (1994).

Stein, J. How institutions learn: A socio-cognitive perspective. *Journal of Economic Issues*, 31(3) (1997).

Steinmueller, W. E. Do information and communication technologies facilitate "codification" of knowledge. *3rd TIPIK Workshop*, April, Strasbourg (1999).

Stinchcombe, A. L. *Information and Organizations.* Berkeley: University of California Press (1990).

Swan, P. *Academic Scientist and the Pharmaceutical Industry: Cooperative Research in Twenty-Century America.* Boston, MA: John Hopkins University Press (1989).

Tagliagambe, S., & Usai, G. *L'impresa tra ipotesi, miti e realtà.* Torino: ISEDI (1994).

Taylor, S. E., & Fiske, S. T. Salience, attention, and attribution: Top of the head phenomena. *Advances in Experimantal Social Psychology*, 11 (1978).

Teece, D. J. Technology transfer by multinational firms: The resouce costs of transferring. *Economic Journal*, 87 (1977).

Teece, D. J. Economies of scope and the scope of the enterprise. *Journal of Economic Behavior and Organization*, 1(3) (1980).

Teece, D. J. Profiting from technological innovation: Implications for integration, collaboraboration, and public policy. *Research Policy*, 15(6) (1986).

Teece, D. J. Competition, cooperation, and innovation: Organizational arrangements for regimes of rapid technological progress. *Journal of Economic Behaviour and Organization*, 18(1) (1992).

Teece, D. J. Capturing value from knowledge assets: The new economy, markets for know-how, and intangible assets. *California Management Review*, 40(3) (1998).

Teece, D. J., Pisano, G., & Shuen, A. Dynamic capabilities and strategic management. *Strategic Management Journal*, 18 (1997).

Thomke, S., & Kuemmerle, W. Asset accumulation, interdependence and technological change: Evidence from pharmaceutical drug discovery. *Strategic Management Journal*, 23(7) (2002).

Toffler, A. *Powershift: Knowledge, Wealth and Violence at the Edge of the 21st Century*. New York: Bantam Books (1990).

Tripsas, M. Surviving radical technological change through dynamic capability: Evidence form the typesetter industry. *Industrial and Corporate Change*, 6 (1997).

Usai, G. L'impresa nel pensiero sistemico-cibernetico, in Caselli L. (a cura di), *Le parole dell'impresa*. Milano: FrancoAngeli (1995).

Van Der Sijde, P. C., & Ridder, A. *Commercializing Knowledge*. Twente University Press (2000).

Veugelers, R. Internal R&D expenditures and external technology sourcing. *Research Policy* (1997).

Vicari, S. *L'impresa vivente*. Milano: Etaslibri (1991).

Vogelaar, G. *Business & Science Park Enschede*. Twente University Press (2000).

Von Hippel, E. *The Sources of Innovation*. Oxford: Oxford University Press (trad. It., 1990, Le fonti dell'innovazione. Milano: McGraw-Hill) (1988).

Von Hippel, E. Economics of product development by users: The impact of 'Sticky' local information. *Management Science*, 44(5) (1998).

Warren, K. The dynamics of rivalry. *Business Strategy Review*, 10(4) (1999a).

Warren, K. The dynamics of strategy. *Business Strategy Review*, 10(3) (1999b).

Warren, K. *Competitive Strategy Dynamics*. Wiley (2002).

Weick, K. E., & Bougon, M. G. Organizations as cognitive maps. In Sims, H. P. Ir., & Gioia, Dennis A. (Associates/Eds.) *The Thinking Organization*. San Francisco: Jossey-Bass (1986).

Weick, K. E. *The Social Psychology of Organizing*. Second Edition, New York: Random House (1979).

Wenger, E. *Communities of Practice*. Cambridge: Cambridge University Press (1998).

Wernerfelt, B. The resource-based view of the firm: Ten years after. *Strategic Management Journal*, 16 (1995).

Whittington, R. The changing structure of R&D: From centralization to fragmentation. In Loveridge, R., & Pitt, M. (Eds.) *The Strategic Management of Technological Innovation*. London: Wiley (1990).

Williamson, O. E. *The Economic Institutions of Capitalism*. New York: Free Press (1985).

Williamson, O. E. *Le istituzioni economiche del capitalismo*. Milano: FrancoAngeli (1987).

Williamson O. E. *The Mechanisms of Governance*. New York: Oxford University Press (1996).

Winter, S. Knowledge and competence as strategic asset, In Teece, D. (Ed.) *The Competitive Challenge: Strategies for Industrial Innovation and Renewal*, Cambridge, MA: Ballinger (1987).

Wofford, J. C. An examination of the cognitive processes used to handle employee job problems. *Academy of Management Journal*, 37(1) (1994).

Zahra, S. A., & George, G. Absorptive capacity: A review, reconcepualisation, and extension. *Academy of Management Review*, 27(2) (2002).

Zucker, L. G., Darby, M. R., & Armstrong, J. S. Commercializing knowledge: University science, knowledge capture, and firm performance in bio-technology. *Management Science*, 48 (2002)

Chapter 7

What Open Innovation Is: Local Search, Technological Boundaries and Sustainable Performance in Biopharmaceutical Experimentation

Maria Rosaria Della Peruta

As is known, in the resource-based view, businesses differ in their resource positions and this heterogeneity of resources is a source of performance differences between businesses (Barney, 1991; Peteraf, 1993).

Such an approach leads to a rethinking of the traditional strategic theory, which, in agreement with neo-classical economics, hypothesized the presence of homogeneous businesses from the point of view of endowment of production factors, because of both the effective similarity of the factors and the mobility of the same and therefore their possible transferability (Wernerfelt, 1984). Other scholars, on the contrary, have stressed a substantial diversity in the endowment of resources, hypothesizing a heterogeneity, difficult to change over time (Helfat, 1994; Knott, 2003; Henderson and Cockburn, 1994; Iansiti and Clark, 1994; Berman, Down, and Hill, 2002; Knott, 2003; Zott, 2003) but have provided only partial indications on how these heterogeneous resource positions come about. While each of the investigated factors, such as initial resource endowment and important investments (prior commitment) (Eisenhardt and Schoonhoven, 1990; Helfat and Lieberman, 2002), timing (Stinchcombe, 1965; Zott, 2003), and management capacity (Knott, 2003), gives interesting explanations of the heterogeneity, none of them offers solutions to the question of how these positional resources were initially acquired.

The question is where the heterogeneity of resources originates. Drawing from evolutionary theory we may identify the originating sources of the heterogeneity of resources as something created in response to idiosyncratic situations (Holland, 1975; Nelson and Winter, 1982).

Businesses respond to opportunities and to idiosyncratic problems that face them by undertaking new search paths, the creation of which represents the keystone of heterogeneity of resources. While different authors have dedicated themselves to examining how "inertia" (Fredrickson and Iaquinto, 1989; Helfat, 1994) and "momentum" (Miller and Friesen, 1980; Amburgey and Miner, 1992) support the change along pre-existing paths (path-deepening search),[1] much less is known about how these emerge in the first place (path-creating search) and whether, once created (created path), these can favor differences in sustainable performance between businesses. This creates an interest in identifying the determinants of the path-creating search and understanding the resulting implications for resource performance.

The objective is to go beyond the study of the differences between businesses in terms of performance in order not only to identify the factors that are correlated with superior performance but also to try to explore the origins and dynamics of their adoption (Cockburn, Henderson and Stern, 2000, p. 1124) and understand the processes through which businesses create competitive advantage.

In this light the generation of new knowledge, that is, the carrying out of learning processes, is fundamental for creating value and takes on a path-dependent nature, in that it is based on processes that activate search routines of a local type, to exploit the latent potential in the knowledge that already exists. In a particular way, search activity also means a problem regarding the selection of the learning sources of the organization. R&D represents a key resource in the process of learning (cfr. Cohen and Levinthal, 1990; Pisano, 1994; Hoang and Rothaermel, 2010); its primary function is to generate new knowledge by "recombining" existing knowledge. Different choices regarding the knowledge to use in the recombination can lead to different paths and to different technological capacities and, as a consequence, to different performances.

Starting from these premises, an attempt has been made to investigate how the phenomenon is manifested within the pharmaceutical sector, where the generation of new molecules able to treat socially widespread pathologies, if adequately protected by patents that temporarily limit competitors with respect to the new drug, has always represented one of the most efficient ways of sustaining a competitive advantage.

Pharmaceutical business frequently commits errors[2] (further demonstrating the evolutionary nature of the process of creation of resources), in some cases not reaching or even exceeding the most productive levels in the activity of R&D, which are linked to the economies of scale, which itself can be an important source of heterogeneity of resources (Ahuja and Katila, 2004): businesses differ in the ways they adopt to resolve the problems in the development of products, and such differences can lead to variations in performance (Dowling and Helm, 2006).

LOCAL SEARCH AND TRADITIONAL DRUG DISCOVERY

Competencies regarding the processes of research and development are considered relevant characteristics that help in differentiating successful businesses from others (Bettis and Hitt, 1995; Teece, 1982). Hence, an understanding of the evolution of capabilities allows identification of performance differentials between businesses (Nelson, 1991; Tsai, 2001). In fact, if capability is defined as specialization or concentration of the activity of a business within certain areas (Selznick, 1957; Smith, 1776), exploration of the evolution of these specializations necessitates investigation into how these specializations emerge in certain technological areas within the organization.

R&D activity involves recombinant processes on the part of the inventors, and the different knowledge resources of a business represent different alternatives available for the business in the recombination process (Dosi, 1982). Consequently, the choice of knowledge asset in the recombination process has an important influence on the course of specialization in R&D and can lead to differences in business capabilities over time. Because of limited resources, businesses cannot follow all the alternative paths available while they guide the R&D processes. The effects of limited resources, of limited rationality, and of incomplete information lead businesses to direct their forces in R&D toward some areas at the expense of others. Some solutions constitute the foundation of the future development of knowledge while other solutions become dead ends (Podolny and Stuart, 1995).

The activity of knowledge generation (searching for knowledge) created within a business, counterposing that created outside the organizational boundaries, is a habitual and routine "response" of inventors.

This behavior implies that inventors alter one component at a time, either reconfiguring it in relation to other components or substituting it with other components, experimenting innovative solutions in an incremental way (local search):[3] in this way research activities are strongly connected to precedents and give value to experience achieved with technologies that are already widely developed.

The traditional procedure for the creation of a new drug (traditional drug discovery), for example, which is founded on the principles of chemical synthesis, followed this model; specifically, after having identified the most promising molecules (candidates) through casual research or the screening of thousands of compounds (luck or high-throughput screening), researchers set up extended experimentation and verified their efficacy in the field under differing conditions (Drews, 2000).

The principal conditions of research in the pharmaceutical sector have always made the discovery of a promising drug an event of a casual nature.

As shown by Henderson (1994), in these circumstances the competencies that businesses found themselves governing in order to obtain product innovation were prevalently connected to the knowledge of chemical synthesis and pharmacology.

Inventors had an extremely limited understanding of the elements they had to recombine and the ways in which these elements would interact—due to limits of knowledge and to fundamental uncertainty (March and Simon, 1958; Nelson, 1982; Vincenti, 1990), so the best strategy for discovering a potentially effective drug was based on the capacity of managing a wide-scale operation able to examine thousands of compounds.

In this way inventors accumulated experience and had strong traditions in a specific therapeutic area, developing greater knowledge of the molecules that allowed them to "invent" with greater reliability, avoiding, that is, molecules that had not worked in the past and choosing those that demonstrated, *in vivo,* the same therapeutic action as other, previously tested, drugs present on the market (Vincenti, 1990).

Therefore, the innovative process in the pharmaceutical sector concentrated the experimentation of new drugs on the capacity of interaction of new molecules, preferring the "familiar" combinations and precluding the possibility of investigating potentially more useful solutions that had unknown or poorly known mechanisms of action (March, 1991; Fleming, 2001).[4]

Science Search and Rational Drug Design (Aa Scientific Approach to Pharmaceutical Research)

Since the 1970s, enormous advances in the pharmaceutical research technology and consequent developments in scientific disciplines, thanks to the opportunity of observing the properties of materials at the microscopic level, have represented idiosyncratic situations (Ahuja and Katila, 2004), which pushed the pharmaceutical companies to look beyond local technology search, crossing the technological barrier to develop new scientific knowledge and to have access to a more heterogeneous resource base for enriching and governing the process of invention.[5]

Science can improve the inventor's comprehension of the cause–effect relationships and in this way can help in identifying combinations of elements that may have an advantageous result (Freeman and Soete, 1997; Cockburn et al., 2000).

Scientific knowledge can lead to very different types of research: effectively, it provides inventors the equivalent of a map, a stylized representation of their area of research.

Science tries to explain why phenomena occur and provides instruments for predicting the results of untried experiments and the utility of previously uncombined configurations of technological components. Having knowledge of the fundamental problem—a map—probably modifies the research process in many complementary ways. That could take inventors directly to appropriate combinations of components for resolving a particular technical problem. Even though no clear connection is evident, science can, ultimately, increase the efficacy of research, with early identification of useful directions of research and the formation of vague notions of what is possible to develop and test. Theoretical knowledge of the basic properties of technological components and their interaction can facilitate efficient research: in fact, scientific theories do not predict all the properties and interactions possible but allow evaluation of the offline alternatives, taking researchers to new promising inventions in a more efficient way (Nelson, 1982).

In other words, science can tell inventors how to avoid wasted effort. It also improves the efficacy of research, impeding inventors from wasting resources even when there is no accurate or complete knowledge of the problem, in that a predictive model can still alter the research process in a useful way. On the other hand, suggesting that a solution could be found from a theoretical point of view, science can encourage inventors to continue to work with an apparently sterile set of components.[6]

To use a metaphor now common in pharmaceutical laboratories, if the action of drugs on the receptors that are on the surface of target cells can be considered analogous to that of a key inserted into a lock, then thanks to the scientific discoveries mentioned, today's knowledge of how locks are made has increased considerably, and it shall become a lot easier to design adequate keys (Henderson, 1994). In this new context, therefore, instead of operating with an essentially inductive approach (that is, which emerges from experiments in the laboratory), which is founded on a high-dimension casual process necessary to overcome the changeable statistical threshold of uncertainty, work is mainly at a deductive level (that is, "at the table"), designing new drugs that, in any case, require subsequent extensive *in vivo* and *in vitro* experimentation.

In this way we have seen the technique of "guided and rational research" (rational drug design) become established, with which researchers develop analytic knowledge about the pathologies and work backward to identify drugs that inhibit it.

Innovative processes often start with known final results and try to find the unknown starting positions that produce the desired behavior. The term "symmetry breaking" stresses that such final conditions cannot be found directly using scientific knowledge, which can be used only to go in the opposite direction, from initially known conditions to unknown final results. Effectively, science is a "one-way street" and technological change goes in the opposite direction. The tacit knowledge of the inventors is used to "examine" how the problem confronted correlates with similar problems confronted in the past (Vincenti, 1990; Nightingale, 1998). If these similar problems have a known solution, one may extrapolate a similar, uncertain situation, making the problem more specific (Vincenti, 1990). This knowledge of what solutions are appropriate for what problems is technology specific and generates ways of resolving problems, called technological traditions.

Once technological traditions define how a problem is to be resolved and research has generated the performance criteria of the problem, inventors use previous experience to suggest a feasible "first cut" solution. This "first cut" solution is the set of initial conditions that are thought to come near to the final result required. This new knowledge is then used to modify the solutions for the next round of experimentation.

During this cycle of comprehension, modification, and experimentation of uncertain solutions, scientific knowledge is used for understanding and predicting behavioral models, so as to eliminate improbable alternatives and understand how things work: scientific knowledge cannot directly create the desired result, and in order to undertake the vast research in order to understand the phenomenon that the inventors are trying to modify, it resorts to the inventor's tacit knowledge to suggest an uncertain "first cut" solution; once such a solution is tested, the results of the test are used to increase, in terms of variety, the assets of routines known to the pharmaceutical business, adding to its innovative capacity. This increase, in turn, stimulates the push toward technological progress at an aggregate level, and over time this selects businesses unable to sustain growth and attributes particular importance to processes of learning activated by businesses in their strategic interactions.[7]

Without going into the specifics of each technological tradition, we wish to stress how they represent one of the most significant processes that lead to the key role played by research activity in guiding the behavior of pharmaceutical businesses.

Substantially, for pharmaceutical businesses, in a sector dominated by scientific discoveries and/or technological advancement, including those deriving from external sources, learning is believed to be extremely critical for increasing the assets of tacit knowledge, which nourish competencies regarding the strategy and governance of businesses. The different trajectories traced by the learning processes of the single businesses influence innovative opportunities and create idiosyncrasies in their evolution.

Specific Knowledge and Research Performance (Boundary Spanning and Exploration Alliances)

In the context of "science-driven drug discovery" (Cockburn et al., 2000; Nightingale, 2000), the procedures of knowledge management specific to business can have a crucial role in explaining research performance differences (Bierly and Chakrabarti, 1996), particularly between competitors with similar intensities and levels of investment in R&D activity.

The description of the learning process is not restricted to the traditional operational functions carried out by the business, but rather is extended to the process of allocation of resources, the informative process, the problems of incentives, and controls.

The contribution of the new technology "subcomponent" to the performance of the whole system is partly correlated with problems of coordination and control of knowledge within the innovative process.

There are two aspects to analyze. First of all, the complexity of the innovation process means that it is very difficult to understand the relationship between micro-changes and macro-effects. This problem is not specific to the pharmaceutical industry. The introduction of mass production into manufacturing at the start of the plast century required a long period of incremental learning and organizational change before performance improved. Similarly, the pharmaceutical industry may have to wait for some time before optimizing its processes.

Secondly, and more importantly, the quality of the experiments depends on how they modify the tacit knowledge of the technologist or inventor, effects that, even if identifiable, cannot be measured directly. In fact, while it is easy to measure the changes in the number of experiments, the corresponding improvements in learning are more difficult to measure. Since testing the wrong type of compounds is a waste of time, simple quantitative measurements are, in themselves, inefficient, and the correct tradeoff between quality and quantity in the experiments is extremely uncertain.

The possible effect each change, in any part of the innovative process, can have on the rest of the system is uncertain because the correct tacit "sense of similarity" can only be ascertained after the drug has been discovered and has been proven successful medically and commercially.[8]

Uncertainties are pervasive and the key for the performance of the systems involves control and management of such uncertainties; the latter isare intrinsic to the system and regard the subjective quality of the experiments and explains why almost a century separates the introduction of mass production into manufacturing and into R&D.

The pharmaceutical industry is not technologically unsophisticated if compared with the small-arms manufacturer of late plast century; rather it is attempting a much more complex project. What differentiates innovative processes in the pharmaceutical sector, seen as a technical system, is that it lacks a load factor (criterion of imputation) (Hughes, 1983, pp. 218–221), which can be used to objectively measure the contribution of each part of the performance of the whole system. Production engineers have parameters and well-defined repartition bases that can be used to calculate and optimize productivity, while the complexity of drug discovery means that decision making is decentralized to experts whose choices are often based on tacit knowledge rather than on clear parameters. This derives from Bradshaw's paradox applied within the innovative process: you cannot be sure that researchers are looking in the right place until they find what they were looking for. As a consequence, unlike the production system, here it is difficult to correlate the performance measures of the whole system—drugs produced for millions of dollars of R&D investment—with performance measures of the component parts.[9]

Even if the changes happening in the innovative processes cannot remove the inherent uncertainty associated with drug discovery, the efficiency of different organizational and managerial reactions to systemic uncertainties has an important effect on the performance of pharmaceutical businesses.[10]

The main implications regarding local search have not specifically advanced hypotheses that sustain that pharmaceutical businesses can explore particular technologies by integrating developments generated by other businesses. In reality, innumerable empirical studies, in different sectors, attribute organizational boundaries between businesses a role as markers of different types of exploration.

In their sample, Stuart and Podolny (1996) demonstrate that only Matsushita was able to reposition itself technologically, distancing itself from local search and making use of alliances with other businesses that gave it access to different technologies. Similarly, Nagarajan and Mitchell (1998) demonstrate that businesses that want to generate "encompassing" technological change must rely on coordination between businesses through strong interrelationships. These studies suggest that bridging the organizational boundaries of the business also means crossing technological boundaries, constructing a sustainable competitive advantage from the ability of the business to move beyond local search, and reconfiguringe its own knowledge base. Even though local search technologies are primarily attributed to internal organizational limits, this tendency is often seen in relationship mechanisms activated by businesses that interact more frequently with other businesses possessing a similar technological background, like social networks emerge between professionals with common technological interests (von Hippel, 1987), and these external relationships reinforce the internal organizational emphasis toward local search.

On the other hand, empirical studies made on patent data suggest a relationship between technological similarity and knowledge flows (even though they do not immediately identify the contribution of such results to creative processes in business). For example, trying to analyze how businesses undertake joint exploration of different technological solutions, Stuart and Podolny (1996) show the technological landscape through citations of patents and the tendency of businesses to maintain a similar position in this landscape for some ten years. This implies that businesses continue to draw on the knowledge of other businesses technologically more similar to themselves.

Competition dynamics, however, encourage diversification through technological subfields to keep one's own competitive edge: among the largest Japanese semi-conductor businesses, the authors show that Matsushita managed to make a technological transition through the strategic use of alliances.

Though the formation of alliances and the employment of researchers represent some useful mechanisms for acquiring knowledge, as is shown in the literature, can these mechanisms be retained as useful for the acquisition of knowledge on the part of businesses that operate in different technological landscapes? In other words, can businesses use mechanisms to overcome the limitations of "contextually oriented" research?

Certainly, as stressed before, alliances are strictly interrelated to technological contexts. In primis, these move mainly toward these contexts. For example, Mowery et al. (1998) demonstrate that alliances are stipulated with greater probability between businesses that have previously constituted technological alliances. In Stuart and Podolny's study (1996), the "contextual" similarities and the alliances are superimposed, so much so that 9 of the 10 Japanese businesses maintain similar technological positions over time, despite the innumerable alliances.

Is technological similarity necessary to facilitate knowledge flows? As we have noted, few studies support the idea that mechanisms can form a bridge to distant technological contexts (Stuart and Podolny's example of Matsushita, which used alliances for absorbing new technological areas).

Does similarity increase or decrease the level at which these mechanisms can facilitate knowledge flows? There are many reasons why alliances can operate in a more efficient way if the business partners belong to the same technological landscape. The sharing of a common culture, which is more easily achieved at the level of the same technological context, facilitates knowledge flows activated by the alliances, as well as reciprocal interpretation. In the second place, a common context makes the practices and routines of the business partner assimilable, which, besides stabilizing the established relationship for facing and overcoming uncertainties by creating an environment of trust, simplifies the absorption and interpretation of the knowledge reached through the alliance.

On the other hand, sometimes businesses treat alliances as an extension of their internal organization, and have the capacity of exploiting this learning mechanism to face different technological landscapes.

Whereas the formation of alliances between businesses in similar contexts duplicates pre-existing relationships and offers little added value to the business partners—since a business generates access to the knowledge of another business through mechanisms internal to the context, this knowledge has often already been accessible through a large number of informal mechanisms emerging from a similar context—the distance of the technological context can often offer access to important new and exclusive knowledge for innovation by recombination.

The very motivations just mentioned push large pharmaceutical businesses to make an elevated number of alliances with small start-up businesses dedicated to research activity in the biotechnology field, or with university laboratories. Such relationships allow the pharmaceutical companies to explore new approaches to research and new technologies. Above all, they allow such businesses to invest in multiple directions right from the first stages of research, without directly committing themselves in the relative activity, but nonetheless levering access to the competencies of highly specialized partners and to further knowledge developed by these as output of the research carried out. The resulting network of relationships guarantees the pharmaceutical companies the possibility of placing initial bets on a large number of projects, aimed at exploring the different possible evolutionary research paths and, at a later instance, of intensifying or quickly withdrawing their investment, in light of the concrete results obtained by each partner and the direction taken by scientific and technological progress (Koza and Lewin, 1998).

If we observe the process of development of new products as an organizational process of knowledge, then the hope of income translates into the incarnation of expected new knowledge through exploration of a product prototype that can be extended in the process of testing and development. Alternatively, explorative alliances could lead to the codification of knowledge through patents.[11] Most of the previous studies examined the exploration phase of the development process for new products, when *linking firms* make alliances for performance goals (Shan et al., 1994; Deeds and Hill, 1996; Baum et al., 2000; Shipilov and Li 2008) (figure 7.1).

The European Lead Factory is a project started on February 7, 2013, by an international consortium of 30 partners supported by the Innovative Medicines Initiative (IMI), which is the largest public–private partnership in Europe fostering projects based on collaborative pharmaceutical research. The specific aim of the project is to generate "unprecedented opportunities for the discovery of new medicines by providing public partners with an industry-like discovery platform to translate cutting-edge academic research into high-quality drug lead molecules on a scale and speed that was not possible previously."

IMI senior scientific project manager Hugh Laverty declared that the European Lead Factory is supported by IMI because it is consistent with IMI's efforts to encourage a high level of cooperation and knowledge sharing in the field of drug research and development. In particular, the goal is to disclose the chemical libraries of firms in order to make knowledge available to public bodies and small and medium-sized enterprises (SMEs) within Europe. Thus, the project represents a unique experiment to verify if there are alternative ways to perform drug screening. At the same time, it provides academics and SMEs with the possibility of examining a chemical library they cannot find anywhere else. Since IMI will finance selected public projects, it is also expected that the discovery of new drugs will be accelerated by this initiative.

Obviously, the ultimate objective is to provide physicians with the new drugs they require to treat their patients better. With regard to this, collaboration has been a topic widely taken into account so far. Martyn Banks, executive director at Bristol-Myers Squibb, discussed the importance of collaboration during a SLAS2013 panel presentation. In an interview for the SLAS Electronic Laboratory Neighborhood e-zine, Banks stated that the common aim at SLAS2013 was to understand how it is possible to work together, so everybody can benefit from screening and early drug discovery. The manager described this as the ecology of drug discovery, in which a partnership is established between pharmaceutical companies, biotech firms, government, and academics all working together. According to Banks, it is paramount to deal with the problems and difficulties related to the development of such a complex network.

TI Pharma scientific director Ton Rijnders claimed that the idea that gave birth to the European Lead Factory was generated by pharmaceutical companies in the European Federation of Pharmaceutical Industries and Association (EFPIA). These firms believe that this is an incredible opportunity to extend their access to new and innovative chemistry for their markets. They are able to have early access to compounds and new targets, as they benefit from improved screening abilities. According to Rijnders, the chemistry academic groups that take part in this project will actually verify how their chemistry ideas are screened in libraries. This is something that would be quite hard for them to achieve by other means. The academics that bring in screens will have the opportunity of accessing a unique collection of compounds that would have normally been out of their reach, and if they desire so, they will have the possibility of developing their own results on that basis.

TI Pharma is a nonprofit organization with the aim of making innovative pharmaceutical research possible, which may lead to novel drugs that can help improve human health. TI Pharma was chosen to supervise the European Lead

Figure 7.1 European Lead Factory: An Open Innovation Experiment in Drug Discovery

Factory because of its experience in creating and managing international public–private partnerships. Rijnders, who has the prime responsibility in the project for screening, has always sustained open innovation and crowdsourcing as a means to enable companies to benefit from access to combined mutual knowledge. Rijnders believes that efficiency and creativity in marketing drugs can be enhanced by open innovation.

The point of departure for the European Lead Factory was an open call, as it occurs for all IMI projects. Between March and May 2012, those interested in participating were required to submit an Expression of Interest that was later fully reviewed. Those selected as finalists were requested to submit a complete project proposal by September 2012. After the proposals were finally approved, the phase of making offers and negotiating contractual terms was initiated.

As stated by Rijnders, at the beginning, most of the debate was about the way to ensure that the outcomes of the project were sustained by a balanced sharing-and-reward system. The result of this discussion was a plan that met everyone's interest, from academics to pharmaceutical companies, since it guaranteed that project outcomes could always be developed by any target owner. Every partner is always in command with regard to how the project proceeds, irrespective of the decision to be made, whether to file a publication, utilize a compound to develop a new medicine, or increase business partnerships. Certain reference points are previously defined so the entire consortium can profit when the hit becomes commercial.

Rijnders claimed that it was quite difficult for all the consortium members to come to an agreement on this complex matter. IMI has established a set of IP rules to be observed while public money is being spent. The contract was drafted also with the aid of attorneys from partners. Although the issue was troublesome, the solution was attained quite rapidly. Negotiations lasted four months, from October 2012 to January 2013. If all partners, and pharmaceutical companies in particular, had not been committed, the solution would not have been possible. All partners truly want the project to work, and for this reason they found a swift agreement.

How the project works

The European Lead Factory has currently received funds for five years amounting to a total of €200 million. Its aim is to create a library of 500,000 compounds and perform about 48 high-throughput screens (HTS) every year. Half of these will be carried out by pharmaceutical company partners while the remaining half will be chosen from the public sector after competitive calls for proposals have been made.

The initiative is divided into two parts, the Joint European Compound Collection and the European Screening Centre. For the project to be successful, the creation of an outstanding collection of compounds is required beforehand. As part of the agreement, the seven EFPIA firms that take part in the European Lead Factory will provide at least 300,000 compounds they have in their personal chemical libraries. Moreover, participating academic institutions and SMEs will contribute with another 200,000 novel compounds.

Gathering the compounds is described by Rijnders as step one. It is important that in the first stage every company submit their compound sets so

Figure 7.1 (Continued)

the library can be activated. At the end of May 2013, the initial estimate of 300,000 compounds had been surpassed. Then, it is essential to make sure that every compound is present only once, so a check for redundancies has to be performed. Moreover, it is fundamental to verify that no compound is available through commercial sources. Rijnders forecasted that only at the end of summer 2013 it will be possible to complete collection and re-plating. At the same time, recruiting assays from inside the consortium have has begun so the assay development process can proceed. Screens should start running in the second half of 2013.

According to Rijnders, an additional key factor that sustains success is the possibility to work in exceptional facilities, such as Merck sites in Scotland (Newhouse) and The Netherlands (Oss), which have been updated with the most innovative equipment for compounds and screening, respectively. The latter, in particular, now known as Pivot Park, has been re-tooled to comply with the ultra-high-throughput screening (uHTS) requirements of the European Lead Factory, and therefore it is extremely interesting to SLAS members. Over a period of five years, 120 public screens are to be run so the most precise and outstanding technologies had to be implemented. In the short term, the goal is to get from 10 to 15 public screens done by the end of 2013.

The targets

When it came to negotiating the contractual terms of the European Lead Factory, the main issues were ownership and the way targets could be addressed. As Rijnders explained, only the target is shared. The point of arrival is a qualified hit list, meaning a set of compounds that satisfies certain requirements and provides significant outcomes that may be liable to advancement. Each program owner may get a maximum of 50 compounds, so the compound library is not emptied out. In an article published on Nature *on February 7, 2013, it is claimed that the difference between the European Lead Factory and the United States National Institutes of Health Molecular Libraries Program is that in the former, "both the chemicals in the screening library and results from the assays will be proprietary. [European Lead] Factory partners will get first right of refusal in licensing deals."*

The struggle for success

Both Laverty and Rijnders claim that the European Lead Factory is only an experiment, although it has been conceived to drive its members toward success.

Laverty asserts that above all it is important that the project has reached its objectives in providing a unique platform that all partners and public third parties can access. Everyone wishes that target owners can exploit the intellectual property generated by the results of the screens. The project manager believes that it is also desirable that future collaborations based on the factory's output can be forged. In the long run, it would be thrilling if chemistries derived from the results of screens accessed pre-clinical development programs, especially those of public third parties.

Rijnders, for his part, claims that one of the objectives of the project is to improve the utilization of existing libraries within firms, so new chemistry on

Figure 7.1 (Continued)

relevant targets can be generated, and the libraries can be disclosed to competitors. It is also an experiment for academics to verify if they are able to enhance the innovative potential on both the chemistry and the biology side, in order to combine innovative target ideas with chemistry they could not access elsewhere to start up new projects. These new models of open innovation and knowledge sharing are quite different from those examined from a traditional standpoint.

The final objective is to discover new drugs that may improve the health of patients. This goal will not be achieved in with the blink of an eye, but all efforts are addressed to that end.

European Lead Factory members

Pharmaceutical companies

- *Bayer Pharma, Germany*
- *AstraZeneca, Sweden*
- *H. Lundbeck, Denmark*
- *Janssen Pharmaceuticals, a company of Johnson and Johnson, Belgium*
- *Merck, Germany*
- *Sanofi, Germany*
- *UCB Pharma, Belgium*

Universities, research organizations, public bodies, nonprofit groups

- *Foundation Top Institute Pharma (Stichting Top Instituut Pharma), The Netherlands*
- *Leiden University, The Netherlands*
- *Max Planck Gesellschaft zur Förderung der Wissenschaften, Germany*
- *Radboud University Nijmegen, The Netherlands*
- *Stichting Het Nederlands Kanker Instituut, The Netherlands*
- *Technical University of Denmark, Denmark*
- *Universität Duisburg-Essen, Germany*
- *University of Dundee, UK*
- *University of Groningen, The Netherlands*
- *University of Leeds, UK*
- *University of Nottingham, UK*
- *University of Oxford, UK*
- *VU-University Amsterdam, The Netherlands*

SMEs

- *BioCity Scotland, UK*
- *ChemAxon, Hungary*
- *Edelris, France*
- *Gabo:Mi Gesellschaft fur Ablauforganisation:Milliarium, Germany*
- *Lead Discovery Center, Germany*
- *Mercachem, The Netherlands*
- *Pivot Park Screening Centre, The Netherlands*
- *Sygnature Discovery, UK*
- *Syncom, The Netherlands*
- *Taros Chemicals, Germany*

Figure 7.1 (Continued)

FINANCING CHOICES AND PERFORMANCE

The Allocation of Ownership Rights to Maximize Research Efficiency

Aghion and Tirole (1994) set out a theoretical model for analyzing the organization of research and the allocation of ownership of the innovation in an *R&D alliance* between a research business and a client business financing the research. The authors start from the assumption that the *research-intensive* businesses have very scarce financial resources and are limited to the possibilities of borrowing money or of commercializing their own innovations; as a consequence, such businesses turn to a "customer" business, which, while unable to develop the innovation independently, can benefit directly from the innovation of the research-intensive business. The contractual position of the two businesses has an impact on the successive allocation of controlling rights to the innovation.

The Aghion–Tirole model leads to two important reflections for this study. First, to maximize the efficiency of research, and later the creation of common value, the controlling rights should be assigned to the research business in all cases in which the value of the final product depends more on the marginal efficiency of the research work than on the marginal impact of the financial investment. On the other hand, *a cash constraint* on the research business causes an inefficient result due to the financier's advantage of using its contractual power to claim ownership. In fact, research businesses, such as biotechnology ventures, normally face a cash constraint, which limits their contractual power and their partners in the alliance use their own financial power to lower the cost of research and achieve ownership at the expense of the research business.

Previously we argued that because of their limited resources, technological ventures choose to make strategic alliances (exploration and exploitation alliances) whenever they attain new knowledge in any part of the development process of the product, thereby transforming the scientific discoveries from pharmaceutical research into to the commercialization of new drugs. In all cases, as predicted by Aghion and Tirole (1994) and demonstrated empirically by Lerner and Merges (1998) and Lerner et al. (2003; Lerner, 2009), financially constrained businesses tend to cede ownership of their innovation when they enter into alliances.

The authors demonstrate that the number of *control rights* allocated to a biotech business is a function of the financial conditions of the business and the presence of favorable conditions in the marketplace:

in fact they find that biotech businesses with larger incomes in the year before stipulating an alliance are less disposed to cede their control rights and that a survey of incomes serves as an indication of the positions of force of biotech businesses that negotiate control rights. Lerner and Meges conclude that the greatest effect of the allocation of control rights, at least in technological alliances, depends on the financial condition of the R&D firm rather than on conditions regarding the maximalization of the common result (Lerner and Merges, 1998). Such observations imply that alliances, as incomplete contracts, are characterized by contractual risks that seem to be exacerbated by the financial limits of the research business.

Unlike Aghion and Tirole, who do not explicitly consider the role of public financing, Lerner et al. (2003) demonstrate the previous theoretical claims empirically,[12] exploring the role played by the availability of public financing with regard to the contractual power of intensive research businesses and the subsequent allocation of control rights. The intention is to verify whether such alliances attribute different allocations of control rights in periods characterized by differing availability of public financing; in particular, the authors investigate whether *success rates*[13] differ for businesses that stipulate agreements in periods of scarce financing and whether these agreements are renegotiated at a later date. The theory suggests that the agreements stipulated in periods with limited availability of external financing probably maximize innovative output to a lesser degree. Lernet et al. (2003) find that in periods of limited public financing, small biotech businesses seek out large pharmaceutical businesses for financing alliances. In these agreements, the biotech businesses are more disposed to cede a greater number of control rights to the financing firm (pharmaceutical business): empirical evidence regarding the number of agreements that are renegotiated when the relative contractual position of the intensive-research business improves as a result of increases in the availability of external financing also supports this hypothesis.

Information Asymmetry, Products in Development, and Financial Limitations

Technological alliances generally possess an information advantage in assessing the quality of their development projects (Lerner et al., 2003). In particular, a technological alliance often pursues several concurrent projects and, because of its established familiarity with projects that foresee distant time horizons, it is reasonably able to understand which projects have better prospects for development.

In the first stages of the development of a product in the biopharmaceuticals sector, because of the problems of information asymmetry, the incumbent business that tries to commercialize a project offered by a technological alliance is not able to judge whether the offered project for collaboration is promising or not, and very often a good project is discarded to avoid being sold "a lemon." Akerlof's model regarding used cars has shown that information asymmetry can lead to a perverse effect where sellers offer only *lemons* since the price offered by buyers is lower than the value at which the sellers are prepared to relinquish a car in good condition.

Applying this analysis to the market of collaborative know-how in the biotechnology setting, Pisano (1997) found empirical evidence for the problem of lemons in the market of drug development projects reaching the phase of clinical trials. When a new venture grows and matures internal resources to finance its wide range of projects, it tends to develop them in-house rather than through alliances with partners well established in the sector. Thus, the lemons problem, together with the downward pressure on market prices for collaborative know-how due to the weak contractual position of the less-than-numerous, under-financed research businesses, creates a situation in which even a well-financed research business has an incentive to keep projects, despite its ability to improve its contractual position.

The gradual trend of substituting external resources with internal resources, when the venture technology is developed, is also supported by theoretical predictions of an optimal model for the structure of capital (Myers, 1984; Myers and Mjluf, 1984). Technological ventures generally find themselves struggling to acquire resources from skeptical investors who have difficulty in judging the quality of specific projects. According to the *pecking order theory* for the optimal structure of capital, the preferences of a business in obtaining capital will follow a hierarchy, with a predilection for internal sources over external sources of capital because of the information asymmetry. This hypothesis is based on the observations in Myers and Mjluf (1984), who reach the conclusion that problems regarding objective risk hinder external financing of risky economic activities. In fact, since those who operate outside recognize an information asymmetry in the fact that whoever operates inside has superior knowledge of the investment opportunity, the cost of financial investment through external funding is higher than the cost through internal financing.

In particular, the propensity for internal resources to finance the most promising projects is based on the cost levels associated with the information asymmetry between the business and the providers of external resources (Myers and Majuluf, 1984; Shyam-Sunder and Myers, 1999) and the risks of expropriation of the knowledge due to opportunist action on the part of the partners (Williamson, 1985).

Initially, new technological ventures with low resource endowments and private information on the projects of value are forced to exchange the ownership of their own projects at a sub-optimal price (Aghion and Tirole, 1994; Lerner and Merges, 1998). This implies that businesses with limited resources are forced to be more than trusting of the external resources to be invested for the development process of new products, running the risk of seeing expropriation of their own *core knowledge assets.*

Corporate Venture Capital, New Ventures, and Performance

It must now be examined to what extent the sector in general and its technological characteristics can lead to the decision of pursuing innovation through *equity investment* in new businesses. On this question, we claim that the marginal benefit of *corporate venture capital* (CVC) is greater in the pharmaceutical sector, which has consistent technological opportunities, has weak protection of intellectual property (in particular that regarding patents), and where the complementary competencies (for example, production and distribution) are important for being able to set aside profits for the innovations. We shall take these aspects into consideration one at a time.

As propounded theoretically, the marginal benefit of CVC should be higher in sectors that have greater technological opportunities; that is, when "the technical development, evaluated on the basis of the current price of production factors, costs less" (Cohen, 1995). The level of such opportunities is influenced, above all, by the progress made in the field of the scientific base and in related technological sectors (Klevorick, Levin, Nelson and Winter, 1995). In particular, Klevorick et al. (1995) claim that sectors differ to a significant extent according to their own levels of technological opportunity and that where they have greater technological opportunity it is probable that entrepreneurs are able to identify new innovations of value, and thanks to these remunerative opportunities, they can start up new businesses (Shane, 2001a). In the presence of a pool of highly innovative businesses, the marginal benefit of CVC increases in proportion to internal R&D. In practice, this means that with the increase in new ideas (and therefore the income predicted as a result of the increasing research workload), it is probable that researchers decide to leave the company for which they are working to start up their own businesses. Hence, if on the one hand the costs increase for guaranteeing the presence of qualified researchers in their own internal laboratories, on the other, there is a mean overall increase in the possibility of acquiring knowledge through investments in new businesses.

The relative advantage of CVC, compared to the costs of internal R&D, is all the more marked if the protection of intellectual property is weaker.[14] In circumstances such as these, CVC constitutes an efficient channel for learning from quality businesses (Dushnitsky and Lenox, 2002).

In the absence of legal protection of a patent, a new business can count on only its secrecy to defend its intellectual property. CVC investment offers a way of penetrating the network of secrecy. Usually, the investing companies are part of the board of directors and, in many cases, activate collaboration programs with other businesses with the scope of stimulating dialogue among researchers of the latter and their own scientists.

When the protection of patents is weak, businesses may not be able to impede investors from appropriating fundamental knowledge. The large pharmaceutical companies have the resources necessary for taking legal action and facing other types of challenge that threaten their patents (Somaya, 2002): besides, they possess those complementary competencies in the field of research, production, and distribution channels that allow them to speculate to about their own advantage. On the contrary, new businesses could consider it expensive to obtain patents for their own technologies and lack the necessary funds for attempting to undertake effective legal protection. As a result, when the protection of patents is weak, the business may not be able to impede the divulgation of knowledge in favor of the investors and as a consequence the incumbent companies have more incentive to choose CVC funding rather than internal R&D.

This argument gives rise to an interesting controversy. If the contribution offered to the innovative production of the business by the CVC investment is directly connected to the quality of the business in which it is investing, high-quality ventures can refuse corporate investors so as to prevent the divulgation ofdivulging their own precious knowledge (especially when they are in a regime with weak IP).

Gans and Stern (2000) claim that an alliance with an established business leads the new business to benefits, which, in some circumstances, may compensate the cost of expropriation due to the divulgation of the innovation. Such benefits include complementary competencies like production and distribution that new businesses often lack.[15]

In the pharmaceutical sector, entrepreneurial initiatives can profit from the support of CVC not only through the availability of funds and the improvement in their own business reputation, but also through real progress in R&D and the operations of production, commercialization, and distribution. In the first place, the CVC investor can provide services with added value, similar to those offered by Quality VC funds (Block and MacMillan, 1993; Hsu, 2002). In the second place, they can extend the array of exclusive services that capitalize business resources. For example, the pharmaceutical companies generally offer the right to use complementary resources to the business itself, like laboratories, access to a network of suppliers and clients, immediately available *beta sites*, and access to national and international distribution channels (Acs, Morck, Shaver and Yeung, 1997; Maula and Murray, 2002; Pisano, 1991; Teece, 1986). Finally, the fact that a focal venture is chosen by an incumbent business constitutes a guarantee for third parties and/or the financial markets (Stuart, Hoang and Hybels, 1999). These observations regarding the pharmaceutical sector agree with the conclusions reached by Maula and Lurray (2001), who show that businesses co-financed by CVC programs obtain higher evaluations compared to those financed exclusively by VC projects.[16]

Chapter Summary

The acquisition of technological capacity, on the part of the pharmaceutical businesses that combine and reconfigure resources and capabilities to fuel innovative continuity, shows that the “emergence” of resources is an evolutionary process: businesses experiment to find the correct investment following profound changes in science (in particular in biology, molecular biology, biochemistry, and advances in the single therapeutic areas with regard to pathologies that afflict the human body) and in technologies (essentially those regarding chemical synthesis, genetic engineering, combinational chemistry, high-throughput screening). Pharmaceutical businesses invest in multiple directions right from the first stages of research, without directly committing themselves in the relative activity, but nonetheless levering access to the competencies of highly specialized partners and to further knowledge developed by these as output of the research carried out. The resulting network of relationships guarantees the pharmaceutical companies the possibility of placing initial bets on a high number of projects, aimed at exploring the different possible evolutionary research paths, and, at a later instance, of intensifying or quickly withdrawing their investment, in light of the concrete results obtained by each partner and the direction taken by scientific and technological progress

NOTES

1. For authors who have analyzed the behavior of businesses in experimenting with new solutions (search behavior), businesses that resist change (inertia) and, once the change has started, take to changing in the same direction (momentum).
2. The principle on which learning is based is that of the creation of errors. In fact, it was errors produced over the course of operations that favored the generation of diverse varieties in every form of learning, since the errors feed the formation of new cognitive schemes with respect to those already consolidated and facilitate, therefore, the formation of new knowledge (Vicari and Trailo, 1998; Vicari, 1998). In this view, the generation of knowledge requires putting continuous processes of experimentation and variation into action (Levitt and March, 1988; Warglien, 1990; Vicari and von Krogh, 1992).
3. Evolutionary search has emphasized the role of local search as the behavior of those businesses that look for solutions in the "neighborhood" of their current experience and knowledge (Stuart and Podolny, 1996). Evolutionary theory has built this concept on the contributions of March and Simon (1958) and Nelson and Winter (1982). Empirical evidence covalidates the trends of businesses toward local search. Helfat (1994) has demonstrated that, for petroleum businesses, spending on research and development on the various technologies varied little from year to year. Recently, Martin and Mitchell (1998) have shown that local search leads most of the incumbents in the product markets to introduce designs that are similar to those incorporated in their existing products. Stuart and Podolny (1996) have shown, for large semi-conductor businesses, how the activity of patenting has tended to concentrate the efforts in the technological domain where they had already taken patents in the previous years.
4. Empirical evidence regarding the pharmaceutical sector highlights that businesses normally focus the activity of generation of new knowledge (exploration) on strictly correlated technological domains, identifying their relative boundaries; the business focuses research activity for innovative solutions on similar technologies, creates innovations of an incremental type, and becomes more expert in its current domain. Such behavior makes the businesses able, over time, to consolidate what Rosenkopf and Nerkar (2001) call "first-order

competence." This accumulated experience is considered to be a distinctive competence if it is superior to that matured by competitors and leads to competitive advantage. The focus that supports such first-order competence can lead the business to develop "core rigidities" (Leonard-Barton, 1995) or to fall into a "competency trap" (Levitt and March, 1988) and suggests that businesses concentrate the explorative effort on the internal development of research activity. Empirical evidence, in particular taken from the biotechnology sector (Sorensen and Stuart, 2000), indicates that, even though greater innovation is associated with greater levels of trust in one's own prior developments, this innovation is less relevant for the other members of the technological community and is a guarantee of obsolescence: exploratory activity (search) is located, and therefore limited, within the technological and organizational boundaries (Lahiri and Narayanan, 2013).

5. The classical literature has often stressed that, on the part of businesses, the activity of problem solving in the development of new products is confined to the area neighboring their current research activity and eventually leads to convergence with previous approaches to research (Cyert and March, 1963; Stuart and Podolny, 1996). More recent literature suggests that many successful businesses go beyond the local search, crossing the technological boundaries to develop new knowledge (Nelson and Winter, 1982) and to differentiate their resource position (Rosenkopf and Nerkar, 2001; Hargadon, 2003). From the problem-solving perspective, which drives the activity of product innovation, businesses will expand scientific research (*science search*) when the areas of the current technologies where they are operating reach their limits. Often innovation occurs through the combination and the recombination of existing elements into new artifacts (Hargadon and Sutton, 1997; Kim and Kogut, 1996; Fleming, 2001). The exploitation (*exhaustion*) of the technological domain becomes an opportunity for conditioning resource heterogeneity. While some scholars have documented that businesses are reluctant to dedicate resources for science search (e.g., Henderson, 1994), the idiosyncratic situations (as described above) can provide opportunities to justify the development of scientific activity; businesses that work in *well-exploited technological domains*, that is, domains where the current efforts in product innovation are based on a lot of previous inventions, are, with great probability, seeking a *science base* in a more intense and advanced way so as to have access to a more heterogeneous *resource base*.
6. Take the case of Prozac. Bryan Molloy started research for Eli Lilly with *local search* activity, deriving hundreds of *benadryl* compounds. They all failed. His colleague David Wong continued the research because he believed in a theory that by blocking the absorption of

serotonin, a property of the *benadryl* derivatives, it was possible to cure depression. Using a new experimental technique that allowed a more accurate evaluation, Wong tested the same compounds as Molloy and demonstrated the efficacy of one of them, *fluoxetine* (the chemical name of Prozac).

7. In particular, the modern approach to pharmaceutical research, based on the investigation of the mechanisms of action, can lead to the identification of active molecules for other purposes than the therapeutical area identified as characterizing the portfolio of discoveries of the business. As a consequence, the area of application of the identified compounds could be different to that hypothesized in the initial phase of selection. This is not pure chance, but the fact that the research proceeds by identification of particularly interesting molecules from a pharmacological point of view because of their interaction with a specific target molecules, which in turn can be in more than one physiopathological process. Think of the case of Viagra, a drug developed for cardiovascular applications, which finds additional uses with respect to the original intentions of Pfizer. In the situation where the pharmaceutical business has an awareness of having an innovative compound for a therapeutical area in hand, but where it does not have sufficient capabilities, the identified compound can become the subject of negotiations for *licensing out*.
8. Improvements in the tradeoff between quantity and quality could lead to a compound that binds very solidly to a protein; however, it shall not be successful as a drug if the protein has no effect on the disease.
9. It must be added that, though businesses can more easily recur to the production of new information, since the associated costs of following the technological revolution, for example in the field of bioinformatics, are radically lower, it is just as true that systematic recourse to inductive learning generates a waste of resources, which raises a problem of the costs and risks connected to the learning process.
10. In the short period, the ability of large pharmaceutical businesses to transform the potential economies of scale into real cost advantages shall depend on the ability of organizing and managing the information flows between the *research tasks* (Hopkins, 1998). On the other hand, the literature on innovation in complex technologies would suggest that moving toward *large-scale improvements*, while possible, shall be extremely difficult (Hughes, 1983; Nightingale and Poll, 2000). In choosing the right organizational architecture, the use of information technology to connect *research tasks* does not point toward rigid structures, but rather toward the construction of an organizational flexibility, a diversity of technological options and a "*muddling through*" approach, so as to confront the technological uncertainties (Stirling, 1998).
11. In the biotechnology industry, for example, research collaboration is motivated by a desire to acquire base knowledge, which can be used

to create new molecular entities, which can themselves become part of the development and the process: exploration alliances "predict" the development of products (Rothaermel and Deeds, 2004, p. 204). An example is given in the collaboration of the business Biogen and the University of Zurich. Their cooperation led to the discovery of Intron A, the primary product for undertaking clinical analysis for the treatment of certain types of leukemia and hepatitis C.

12. Such work is supported by the empirical results of Aghion and Bolton (1992) and Holmstrom and Tirole (1997), which show strong links between the transfer of control rights and the conditions of public equity markets, in terms of availability of public financing.
13. Lerner, Shane and Tsai (2003) measure success by analyzing the progress of products through the various stages of *phase testing*.
14. We define a weak IP regime as one where the struggle of a business to protect its own innovations from the risk of imitation occurs through legal instruments like patents (Cohen, Nelson and Walsh, 2001).
15. Gans, Hsu and Stern (2002), basing themselves on research carried out on more than a hundred entrepreneurial initiatives, have noted that cooperation with incumbent businesses, through patenting, strategic alliances, or outright acquisition is the preferred line of action when the solidly held complementary resources of the incumbent business are crucial for the commercialization of the innovation.
16. It is difficult to evaluate, ex-ante, the real level of the contribution offered by a consolidated business to its own portfolio of ventures. However, we know that the access to complementary resources acquires greater importance for some sectors rather than for others (Cohen, Nelson and Walsh, 2001). For example, complementary resources are held more solidly by incumbent business of the chemicals and pharmaceuticals sectors where the production and distribution on a vast scale are relevant. We may, therefore, suppose that in those sectors where the consolidated businesses control complementary resources of crucial importance, entrepreneurs are more disposed to affiliate with company investors. As a consequence, businesses that undertake CVC investments in such industries are able, normally, to profit from *higher quality ventures* and to obtain a greater marginal contribution for their innovative production.

REFERENCES

Acs, Z. J., Morck, R., Shaver, J. M., & Yeung, B. The internationalization of small and medium-sized enterprises: A policy perspective. *Small Business Economics*, 9(1): 7–20 (1997).

Aghion P., & Tirole, J. The management of innovations. *Quarterly Journal of Economics*, 109: 1185–1210 (1994).

Aghion, Philipe & Bolton, P. An incomplete contracts approach to financial contracting. *Review of Economic Studies*, 59(3), 473–494 (1992).

Ahuja, G., & Katila, R. Where do resources come from? The role of idiosyncratic situations. *Strategic Management Journal*, 25: 887–907 (2004).

Amburgey, T., & Miner, A. Strategy momentum: The effects of repetitive, positional, and contextual momentum on merger activity. *Strategic Management Journal*, 13(5): 335–348 (1992).

Barney, J. Firm resources and sustained competitive advantage. *Journal of Management*, 17: 99–120 (1991).

Baum, J. A. C., Calabrese, T., & Silverman, B. S. Don't go it alone: Alliance network composition and startups' performance in Canadian biotechnology. *Strategic Management Journal*, 21(3): 267–294 (2000).

Berman, S., Down, J., & Hill, C. Tacit knowledge as a source of competitive advantage in the national basketball association. *Academy of Management Journal*, 45: 13–31 (2002).

Bettis, R. A., & Hitt, M. A.The new competitive landscape. *Strategic Management Journal*, 16: 7–20 (1995).

Bierly III, P. E., & Chakrabarti, A. K. Technological learning, flexibility, and new product development in the pharmaceutical industry. *IEEE Transactions on Engineering Management*, 43(4): 368–380 (1996).

Block, Z., & MacMillan, I. *Corporate Venturing: Creating New Business Within the Firm*. Harvard Business School Press: Boston, MA (1993).

Cockburn, I., Henderson, R., & Stern, S. Untangling the origins of competitive advantage. Strategic *Management Journal*, Special Issue, 21(10–11): 1123–1145 (2000).

Cohen, W. Empirical studies of innovative activity. In P. Stoneman, ed., *Handbook of the Economics of Innovation and Technical Change*. Basil Blackwell: Oxford, England (1995).

Cohen, W. M., & Levinthal, D. A. Absorptive capacity: A new perspective on learning and innovation. *Administrative Science Quarterly*, 35(1): 128–152 (1990).

Cohen, W., Nelson, R., & Walsh, J. Protecting their intellectual assets: Appropriability conditions and why U.S. manufacturing firms patent (or not). http://papers.nber.org/papers/W7552 (2001).

Cyert, R., & March, J. *A Behavioural Theory of the Firm*. Englewood Cliffs, NJ: Prentice-Hall (1963).

Deeds, D. L., & Hill, C. W. L. Strategic alliances and the rate of new product development: an empirical study of entrepreneurial biotechnology firms. *Journal of Business Venturing*, 11(1): 41–55 (1996).

Dosi, G. Technological paradigms and technological trajectories. *Research Policy*, 11: 147–162 (1982).

Dowling, M., & Helm, R. Product development success through cooperation: A study of entrepreneurial firms. *Technovation*, 26: 483–488 (2006).

Drews, J. Drug discovery: A historical perspective. *Science*, 287: 1960–1964 (2000).

Eisenhardt, K., & Schoonhoven, K. Organizational growth: Linking founding team, strategy, environment, and growth among U.S. semiconductor ventures, 1978–1988. *Administrative Science Quarterly*, 40: 84–110 (1990).

Fleming, L. Recombinant uncertainty in technological search. *Management Science*, 47: 117–132 (2001).

Fredrickson, J., & Iaquinto, A. Inertia and creeping rationality in strategic decision processes. *Academy of Management Journal*, 32: 516–542 (1989).

Freeman, C., & Soete, L. *The Economics of Industrial Innovation*. Cambridge, MA: MIT Press (1997).

Gans, J. S., Hsu, D. H., & Stern, S. When does start-up innovation spur the gale of creative destruction?, *RAND Journal of Economics*, 33(4): 571–586 (2002).

Gans, J. S., & Stern S. The product market and the market for ideas: Commercialization strategies for technology entrepreneurs, Working Paper, Melbourne Business School (2001).

Hargadon, A. *How Breakthroughs Happen: The Surprising Truth about How Companies Innovate*. Boston: Harvard Business School Press (2003).

Hargadon, A. B., & Sutton, R. I. Technology brokering and innovation in a product development firm. *Administrative Science Quarterly*, 42: 716–749 (1997).

Helfat, C. Evolutionary trajectories in petroleum firm R&D. *Management Science*, 40(12): 1720–1747 (1994).

Helfat, C., & Lieberman, M. The birth of capabilities: Market entry and the importance of pre-history. *Industrial and Corporate Change*, 11: 725–760 (2002).

Henderson, R., & Cockburn, I. Measuring competence? Exploring firm effects in pharmaceutical research. *Strategic Management Journal*, Winter Special Issue, 15: 63–84 (1994).

Hoang, H., & Rothaermel, F. T. Leveraging internal and external experience: exploration, exploitation, and R&D project performance. *Strategic Management Journal*, 31: 734–758 (2010).

Holland, J. H. *Adaptation in Natural and Artificial Systems*. Ann Arbor, MI: University of Michigan Press (1975).

Holmström, B., & J. Tirole, Financial intermediation, loanable funds, and the real sector. *The Quarterly Journal of Economics*, 112(3), 663–691 (1997).

Hopkins, M. An examination of technology strategies for the integration of bioinformatics in pharmaceutical R&D processes, unpublished masters dissertation, SPRU, University of Sussex (1998).

Hsu, D. What do entrepreneurs pay for venture capital affiliation? Working Paper, the Wharton School of Business (2002).

Hughes, T. P. *Networks of Power Electrification in Western Society 1880–1930*. Baltimore, MD: Johns Hopkins University Press (1983).

Iansiti, M., & Clark, K. Integration and dynamic capability: Evidence from product development in automobiles and mainframe computers. *Industrial and Corporate Change*, 3: 557–605 (1994).

Kim, D., & Kogut, B. Technological platforms and diversification. *Organization Science*, 7(3): 283–301 (1996).

Klevorick, A., Levin, R., Nelson, R., & Winter, S. On the sources and significance of interindustry differences in technological opportunities. *Research Policy*, 24(2): 185–205 (1995)

Knott, A. M. Persistent heterogeneity and sustainable innovation. *Strategic Management Journal*, 24(8): 687–705 (2003).

Koza, M. P., & Lewin, A. Y. The co-evolution of strategic alliances. *Organization Science*, 9: 255–264 (1998).

Lahiri, N., & Narayanan, S. Vertical integration, innovation, and alliance portfolio size: Implications for firm performance. *Strategic Management Journal*, 34: 1042–1064 (2013).

Leonard-Barton, Dorothy. *Wellsprings of Knowledge: Building and Sustaining the Sources of Innovation*. Boston: Harvard Business School Press (1995).

Lerner, J. The empirical impact of intellectual property rights on innovation: Puzzles and clues. *American Economic Review Papers and Proceedings*, 90(2), 343–348 (2009).

Lerner, J., & Merges, R. P. The control of technology alliances: An empirical analysis of the biotechnology industry. *Journal of Industrial Economics*, 46(2), 125–156 (1998).

Lerner, J., Shane, H., & Tsai, A. Do equity financing cycles matter? Evidence from biotechnology alliances. *Journal of Financial Economics*, 67(3), 411–446 (2003).

Levitt, B., & March, J. G. Organizational learning. *Annual Review of Sociology*, 14: 319–340 (1988).

March, J. G., & Simon, H. *Organizations*. Wiley (1958).

March, J. G. Exploration and exploitation in organizational learning. *Organization Science*, 2: 71–87 (1991).

Markides, C. C., & Williamson, P. J. Corporate diversification and organizational structure: A resource based view. *Academy of Management Journal*, 39(2): 340–367 (1996).

Maula, M., & Murray, G. Corporate venture capital and the creation of US public companies. *Presented at the 20th Annual Conference of The Strategic Management Society* (2000).

Miller, D., & Friesen, P. Momentum and revolution in organizational adaptation. *Academy of Management Journal*, 23: 591–614 (1980).

Mowery, D. C., Oxley, J. E., & Silverman, B. S. Technological overlap and inter-firm cooperation: Implications for the resource based view of the firm. *Research Policy*, 27: 507–523 (1998).

Myers S. C. The capital structure puzzle. *Journal of Finance*, 39(3):575–592 (1984).

Myers, S. C., & Majluf, N. S. Corporate financing and investment decisions when firms have information that investors do not have. *Journal of Financial Economics*, 5: 187–221 (1984).

Nagarajan, A., & Mitchell, W. Evolutionary diffusion: Internal and external methods used to acquire encompassing, complementary, and incremental

technological changes in the lithotripsy industry. *Strategic Management Journal*, 19(11): 1063–1077 (1998).
Nelson, R. Why do firms differ and how does it matter? *Strategic Management Journal*, 12: 61–74 (1991).
Nelson, R., & Winter, S. *An Evolutionary Theory of Economic Change*. Cambridge, MA: Harvard University Press (1982).
Nightingale, P. A cognitive model of innovation. *Research Policy*, 27 (1998).
Nightingale, P. Economies of scale in experimentation: Knowledge and technology in pharmaceutical R&D. *Industrial and Corporate Change*, 9(2) (2000).
Nightingale, P., & Poll, R. Innovation in investment banking: The dynamics of control systems in the Chandlerian firm. *Industrial and Corporate Change*, 9: 113–141 (2000).
Peteraf, M. A. The cornerstones of competitive advantages: A resource-based view. *Strategic Management Journal*, 14: 179–191 (1993).
Pisano, G. The R&D boundaries of the firm: An empirical analysis. *Administrative Science Quarterly*, 35: 153–176 (1990).
Pisano, G. The governance of innovation: Vertical integration and collaborative arrangements in the biotechnology industry. *Research Policy*, 20 237–249 (1991).
Pisano, G. P. Knowledge, integration, and the locus of learning: an empirical analysis of process development. *Strategic Management Journal*, Winter Special Issue 15: 85–100 (1994).
Pisano, G. P. *The Development Factory: Unlocking the Potential of Process Innovation*, Boston, MA: HBS Press (1997).
Podolny, J., & Stuart, T. A role-based ecology of technological change. *American Journal of Sociology*, 100: 1224–1260 (1995).
Rosenkopf, L., & Nerkar, A. Beyond local search: Boundary spanning exploration and impact in the optical disk industry. *Strategic Management Journal*, 22(4): 287–306 (2001).
Rothaermel, F. T., & Deeds, D. L. Exploration and exploitation alliance in biotechnology: A system of new product development. *Strategic Management Journal*: 201–221 (2004).
Selznick, P. *Leadership in administration*. Berkeley: Harper & Row (1957).
Shan, W., Walker, G., & Kogut, B. Interfirm cooperation and startup innovation in the biotechnology industry. *Strategic Management Journal*, 15(5): 387–394 (1994).
Shane, S. Technological opportunities and new firm creation. *Management Science*, 47(2): 205–220 (2001a).
Shane, S. Technology regime and new firm formation. *Management Science*, 47(9): 1173–1190 (2001b).
Shipilov, A. V., & Li, S. X. To have a cake and eat it too? Structural holes' influence on status accumulation and market performance in collaborative networks. *Administrative Science Quarterly*, 53(1): 73–108 (2008).
Shyam-Sunder, L., & Myers, S. C. Testing static tradeoff against pecking order models of capital structure. *Journal of Financial Economics*, 51: 225 (1999).

Smith, A. *The wealth of nation*. Clarendon Press (1776).

Somaya, D. Strategic determinants of decisions not to settle patent litigation. *Strategic Management Journal* (2002)

Sorensen, Jesper B., & Stuart, Toby E. Aging, obsolescence and organizational innovation. *Administrative Science Quarterly*, 45: 81–112 (2000).

Stinchcombe, A. Organizations and social structure. In *Handbook of Organizations*, March, J. G. (Ed.) Chicago, IL: Rand McNally: 142–193 (1965).

Stirling, A. The Economics and analysis of technology diversity. *SPRU Electronic Working Paper Series*, Paper 28, University of Sussex (1998).

Stuart, T., & Podonlny, J. Local search and the evolution of technological capabilities. *Strategic Management Journal*, Special Issue, 17: 21–38 (1996).

Stuart, Toby E., Hoang, Ha & Hybels, Ralph C. Interorganizational endorsements and the performance of entrepreneurial ventures. *Administrative Science Quarterly*, 44: 315–349 (1999).

Teece, D. J. Towards an economic theory of the multiproduct firm. *Journal of Economic Behavior and Organization*, 3(1): 39–63 (1982).

Teece, D. J. Profiting from technological innovation: Implications for integration, collaboration, and public policy. *Research Policy*, 15: 285–305 (1986).

Tsai, W. Knowledge transfer in intraorganizational networks: Effects of network position and absorptive capacity on business unit innovation and performance. *Academy of Management Journal*, 44(5): 996–1004 (2001).

Vicari, S. *La creatività dell'impresa*. Milano: Etaslibri (1998).

Vicari S., & von Krogh G. L'approccio autopoietico all'apprendimento strategico sperimentale, *Economia e Politica Industriale*, 74(76): (1992).

Vicari, S., & Troilo, G. *Errors and learning in Organizations*, In von Krogh., von Roos., & Kleine, D. (Eds.) *Knowing in firms*, Londra: Sage Publication: 204–222 (1998).

Vincenti, W. G. *What Engineers Know and How They Know It*. Baltimore, MD: Johns Hopkins University Press (1990).

von Hippel, E. Cooperation between rivals: Informal know-how trading. *Research Policy*, 16: 291–302 (1987).

Warglien, M. *Innovazione e impresa evolutiva*. Cedam: Padova (1990).

Wernerfelt, B. A resource-based view of the firm. *Strategic Management Journal*, 2(5) (1984).

Williamson, O. *The Economic Institutions of Capitalism*. New York: Free Press (1985).

Zott, C. Dynamic capabilities and the emergence of intraindustry differential firm performance: Insights from a simulation study. *Strategic Management Journal*, 24(2): 97–125 (2003).

Index

absorptive capacity, 121
appropriability, 18, 74, 77

business model, 2–4, 8–9, 16, 32

closed innovation, 2, 16
collective invention, 69, 72–83
colocation, 74, 77
creativity, 19, 35, 52, 55, 59, 60, 64, 65, 91, 108, 109, 112, 167

digital immigrant, 92, 97
digital native, 92, 97, 100, 112
disruptive technologies, 29, 33, 44, 45
drug discovery, 6, 158, 161, 163, 166

E-assessment, 106
E-commerce, 34, 45
education, 31, 54, 57, 59–61, 92, 97, 98, 100–5, 112
Educause Centre for Applied Research (ECAR), 99
European Lead Factory, 166–9
evolutionary economics, 18–19
external knowledge, 3, 12, 18, 19, 41, 83, 93–4, 111, 112
  sourcing, 128
external R&D, 3, 8, 139
external sourcing, 120, 121, 128, 129, 137, 144

Facebook, 38, 39, 99
Firefox, 36

gatekeeper, 11
General Public License (GPL), 82

higher education, 92, 102, 104, 105, 112
hold-up problem, 126, 140

information and communication technologies (ICTs), 96, 97, 98, 99, 100, 101, 104
information technology (IT), 44, 55, 99, 178
innovation, 2, 3–12, 14, 15, 16, 17, 18, 19, 28, 29, 30, 32, 33, 34, 36, 38, 39, 42, 44, 51, 52, 53, 54, 60, 61, 63, 64, 65, 69, 72, 74, 77, 80, 81, 83, 91, 93, 94, 95, 109, 110, 111, 120, 121, 122, 123, 124, 125, 129, 130, 135, 137, 139, 158, 162, 165, 170, 173, 174, 176–7, 178, 179
  management, 2, 7, 33
innovativeness, 17, 93, 94, 110, 111
innovative process, 93, 123, 124, 138, 140, 158, 160, 162, 163
Intel Corporation, 37
intellectual property, 1, 54, 61, 62, 77, 82, 120, 126, 139, 141, 143, 144, 168, 173, 174
  rights, 30, 77, 126, 128
Internet, 20, 31, 32, 34, 35, 36, 37, 58, 81, 92, 98, 99, 101, 102
Internet based tools, 5
Internet Engineering Task Force, 13

invention, 31, 34, 51–66, 69, 71, 73, 75, 78, 126, 127, 159, 177
inventiveness, 10, 52–7, 59, 60, 61, 65, 66
inventor, 40, 53–8, 60, 62, 63, 65, 70, 73, 75, 76, 79, 80, 81, 83, 157, 158, 159, 160, 161, 162

Joint Information Systems Committee (JISC), 98–100

knowledge
  acquisition, 6, 12, 93
  exploitation, 12
  integration, 12
  management, 4, 12, 42, 111, 161
  -sharing network, 95–6
  strategy groups, 6

licensing, 54, 74, 78, 141, 142, 168
  costs, 32
  cross-licensing, 82
  out, 178
Linux, 11, 18
local search, 71, 158, 163, 176, 177

markets of knowledge resources, 120
Microsoft, 35

Net Generation, 92, 96, 97, 98, 100, 101, 106, 112
Nokia, 11
not-invented-here (NIH) syndrome, 12

open innovation, 1–16, 19, 20, 28, 37, 45, 81, 83, 167, 169
openness, 5, 9, 10, 18, 28, 36, 55
open science, 1, 19, 83
open source software, 4, 7, 32, 35
out-licensing, 3, 8

patenting, 17, 54, 61, 62, 63, 64, 65, 176, 179
peer production, 35–6
Procter & Gamble, 7, 14

R&D, 2, 3, 5, 6, 17, 18, 19, 28, 30, 39, 40, 53, 54, 69, 71, 72, 77, 120, 123, 126, 130, 135, 138, 139, 142, 156, 157, 161, 163, 171, 173, 174, 175
  alliances, 170
  capacities, 9
  costs, 9
  management, 7
  resources, 7
RBV, 130, 131, 133, 134
royalties, 54, 127, 142

science, 1, 4, 20, 30, 41, 42, 43, 51, 53, 55, 69, 70, 76, 77, 78, 79, 83, 159, 160, 175
  search, 177
scientific investigation, 56, 139
scientific knowledge, 4, 59–60, 93, 159, 160, 161
scientist, 54, 55, 58, 60, 71, 75, 78, 79, 120, 125, 174
search strategies, 17, 18, 19
social business, 34, 35
social media channels, 34, 35
social media networks, 37, 38, 45
social networks, 40, 94, 99, 100, 101, 107, 163
spinoffs, 8
spinouts, 5
student entrepreneur, 92, 107–12
sustainable development, 64, 65

TCE, 123
technology/technological, 27, 29, 30, 31, 32, 33, 35, 37, 41, 42–4, 51, 53, 55, 61, 73, 74, 77, 80, 82, 96, 98, 99, 100, 104, 120, 122, 127, 128, 136, 137, 139, 141, 142, 143, 144, 159, 162
  community, 80, 177
  invention, 52, 56, 57, 58
  knowledge, 4, 42, 120
  licensing, 12
  management, 29, 33, 42–5
  market, 126, 142

momentum, 80, 81
opportunities, 18, 32, 72, 74, 173
progress, 27, 31, 32, 33, 44, 45, 66, 71, 161, 165, 176
search, 19
Twitter, 38
user knowledge, 17

Web, 0, 2, 101, 103, 106
Web 2, 0 technologies, 101, 103, 104, 105, 106
Wikipedia, 36, 38, 44
World Wide Web, 31, 45

Printed and bound in the United States of America